Math Anxiety — How to Beat It!

How do we conquer uncertainty, insecurity, and anxiety over college mathematics? You can do it, and this book can help.

The author provides various techniques, learning options, and pathways. Students can overcome the barriers that thwart success in mathematics when they prepare for a positive start in college and lay the foundation for success.

Based on interviews with over 50 students, the book develops approaches to address the struggles and success these students shared. Then the author took these ideas and experiences and built a process for overcoming and achieving when studying not only the mathematics many colleges and universities require as a minimum for graduation, but more to encourage reluctant students to look forward to their mathematics courses and even learn to embrace additional ones.

Success breeds interest, and interest breeds success. Math anxiety is based on test anxiety. The book provides proven strategies for conquering test anxiety. It will help find ways to interest students in succeeding in mathematics and assist instructors on pathways to promote student interest, while helping them to overcome the psychological barriers they face. Finally, the author shares how math is employed in the "real world," examining how both STEM and non-STEM students can employ math in their lives and careers. Ultimately, both students and teachers of mathematics will better understand and appreciate the difficulties and how to attack these difficulties to achieve success in college mathematics.

Brian Cafarella, Ph.D. is a mathematics professor at Sinclair Community College in Dayton, Ohio. He has taught a variety of courses ranging from developmental math through pre-calculus. Brian is a past recipient of the Roueche Award for teaching excellence. He is also a past recipient of the Ohio Magazine Award for excellence in education. Brian has published in several peer-reviewed journals. His articles have focused on implementing best practices in developmental math and various math pathways for community college students. Additionally, Brian was the recipient of the Article of the Year Award for his article, "Acceleration and Compression in Developmental Mathematics: Faculty Viewpoints" in the Journal of Developmental Education.

Textbooks in Mathematics

Series editors:
Al Boggess, Kenneth H. Rosen

www.routledge.com/Textbooks-in-Mathematics/book-series/CANDHTE
XBOOMTH

Math Anxiety — How to Beat It!

Brian Cafarella

CRC Press
Taylor & Francis Group
Boca Raton London New York

CRC Press is an imprint of the
Taylor & Francis Group, an **informa** business

A CHAPMAN & HALL BOOK

Designed cover image: Shutterstock

First edition published 2025
by CRC Press
2385 Executive Center Drive, Suite 320, Boca Raton, FL 33431, U.S.A.

and by CRC Press
4 Park Square, Milton Park, Abingdon, Oxon, OX14 4RN

CRC Press is an imprint of Taylor & Francis Group, LLC

ISBN: 9781041013136 (hbk)
ISBN: 9781041010869 (pbk)
ISBN: 9781003614142 (ebk)

DOI: 10.1201/9781003614142

Typeset in Palatino
by Newgen Publishing UK

This is for my beautiful wife, Lisa, and our wonderful son, Gavin.

This is for my loving parents, Margaret and John.

This is for my in-laws, Ethel and Roger, and it was my mother-in-law, Ethel, who gave me the idea for this book!

This is for the late Professor Eric Kraus, an amazing math professor who taught me so much and who is gone too soon.

Contents

Preface

About this Book and How to Use This Book

Who is this Book for?

This book is for people who are contemplating attending community college or college but have a fear or disdain of mathematics. More specifically, this is for people who wish to further their education but feel that math has been and will be a barrier to their success. This may include high school students who do not feel that they are "college material," especially because of math. This book is also for adults who have been away from school for some time and are contemplating going to college but have concerns regarding completing a math class. Additionally, this book is for students who have attempted math in college but have been unsuccessful. This book could be for the non-math major looking to complete a required college-level math class. However, this book can also help a student get started in the STEM (Science, Technology, Engineering, and Mathematics) academic path as well.

Are You Just Going to Lecture Me on What to do in a Math Class?

I share helpful tips based on my years of teaching college math. However, as I discuss throughout the book, as a student, I was not considered "college material," and I dreaded taking a college math class. I also present several testimonials from a diverse group of students who feared and disliked math but eventually completed a college-level math class. I have interviewed these people to gain a perspective of a student's struggles and accomplishments in a math class. These students candidly shared their experiences in math classes and what led them to success.

Should I Read the Book Cover to Cover?

As I wrote the book for the student who is concerned about enrolling in college and attempting a math class, the book does have somewhat of a chronological flow to it. My hope is to help guide the reader from choosing a college and registering for a math class all the way through evaluating his or her performance in a math class. However, it is not imperative to read the book from beginning to end. Certainly, there may be a chapter that stands out that you wish to read first. Again, this book is also for students who have attempted math in college but were unsuccessful; hence, some readers may want to start in different places. Finally, my hope is this book will be a resource to you throughout your endeavor in college math, and therefore, you will be able to use it time and time again.

I Noticed You Discuss some Easy Math Problems. Are You Sure this is to Prepare Us for College Math?

As I will discuss many times in this book, mastering prerequisite skills is an imperative ingredient to success in any math class. Consequently, I discuss several arithmetic and basic algebra topics that are necessary for success in any college-level math class and beyond. I also use various math problems to make points about study tips and test-taking skills. I chose more straightforward topics in the hope that most readers could relate.

I Noticed there are Math Topics that are too Difficult for Me. I Can't Follow those!

I discuss topics throughout arithmetic, basic algebra but also intermediate algebra, college algebra, and some basic statistics for various reasons. As I remind the reader throughout the book, do not worry if you come across a math problem in which you are unfamiliar. You can certainly try to follow along, but if you have not reached the advancement of that problem, it is no big deal. Again, my hope is this book will serve as a continuous resource to you. I want the reader to be able to fall back on this book multiple times when completing a college math class. Consequently, as you gain more mathematical knowledge, the more advanced problems will make more sense.

So, is this Book All Math Talk?

Absolutely not. In addition to how to approach a math class, there are tips on how to make important decisions such as selecting a learning modality (e.g., in person, online), advice for acclimating to college, managing "test anxiety," and math in the real world.

I am still in High School; Can this Book Help Me?

This book can certainly serve students who are currently in high school. Several of the strategies discussed in this book to overcome math anxiety can be extrapolated to students of varying levels, including high school students. Additionally, most of the sample and practice problems in this book are crossover problems that are utilized in both high school and introductory college-level math classes.

Acknowledgments

I would like to thank CRC Press for believing in this project.

I would like to thank all of the student participants over the years who contributed to this work.

1

Why do I Hate Math?

What is Math Anxiety?

There are varying definitions, but I am going to define math anxiety as the fear of practicing mathematics or even approaching or enrolling in a mathematics course. Math anxiety impacts many American students. In fact, about, 93% of adult Americans report experiencing math anxiety. Comparatively, students feel more anxiety toward math than any other discipline (Blazer, 2011). Like any other type of anxiety, math anxiety can be measured on a continuum. Some students simply do not like or even detest math and experience only mild anxiety before an exam. Other students feel major anxiety sitting in a math class attempting to master new math concepts. Then, there are students who avoid going to school because of math, and even the thought of attending a math class causes a panic attack or makes them ill.

Why Should I Listen to This Guy?

If you have made it this far, you are likely still questioning how this book can help you. For your entire academic career, teachers have been telling you what to do. Am I just another teacher who will suggest that you study and complete your work? Yes, I am a math professor and I do have advice for success in math; however, I am also a former student who once struggled in math. Moreover, I was a subpar student for many years, and I was not considered "college material."

My elementary school years were checkered academically. There were times I succeeded, and there were times I failed, and that was the case for my progress in math as well. The reality is that I applied my effort unevenly. Truth be told, I never really saw any value in education. Consequently, so many times I simply daydreamed or "checked out" in class. It seemed like

DOI: 10.1201/9781003614142-1

every year it took a phone call from one of my teachers, a bad report card, a bad parent–teacher conference, or all the above to get me to focus on my studies. This led to stretches of success in my studies and in math. I was able to master concepts such as operations with whole numbers, order of operations, and fractions to some degree. However, the scary stuff really came in middle school.

The math got harder in middle school. After all, seventh- and eighth-grade math typically incorporates the arithmetic concepts that students should have learned in elementary school with more advanced applications, plus some basic algebra. However, math was the least of my problems. Like many adolescents, I struggled with low self-esteem, anxiety, and depression. Keeping on top of my studies became even more inconsequential than it had in the past. All I remember about middle school math was sitting in class feeling absolutely lost. I cannot recount the topics that I was supposed to be learning. Math felt like a foreign language. Incidentally, I was already taking a foreign language (Spanish) at the time and failing that miserably as well. In fact, I was even failing "Keyboarding" class!

In eighth grade, we received numerical grades on our quarterly report cards, and my math grades were 73%, 66%, 62%, 70%, and a 69% on the final exam. I achieved a 68% for the year. I barely passed all of my classes. At one point my father was called in for a conference because my teachers felt that I belonged in special education. At another point my math teacher showed my mother some sloppy and careless work I submitted. My teacher declared, "This hurts when a student submits this kind of work." I was constantly anxious and depressed and saw no value in school, let alone learning math. I emphasize this, as many students struggle with external barriers that negatively impact their success in math. I will also admit that I was lazy and did not want to put forth much effort into my studies. For example, on one social studies homework assignment, after reading a chapter, I was asked to "describe the events of the Battle of Bull Run." Instead of reading the chapter, or even looking up the answer, I answered "It was when they allowed the bulls to run." My teacher was not amused.

Our high school, like many others, utilized tracking. There was an honors track, a college-prep track, a below-average track, and special education. I was placed in the below-average track, as my parents fought to keep me out of special education. However, I did wind up in special education at a future point. This meant I was considered "learning disabled" and would require special services such as attending a resource room for extra help. I continued to struggle in math. Truthfully, math just made me feel stupid. Other students got it, but I could not. I also continued to contend with anxiety and depression. I will not use these issues as an excuse for neglecting my studies. Truthfully, I never saw value in education throughout my K-12 years. However, such issues certainly brought me further away from my studies.

High school continued to be a struggle, but I was able to pass and graduate. As for math? I was able to get by with two years of algebra and a business math class. I marginally passed both algebra classes with "C's." I am still not sure how I passed those classes, as I remember feeling mostly lost. My math class during junior year was a nightmare. My mother worked it out where I could work with a private tutor to complete that math class. With that, junior year ended with a whimper.

My goal throughout high school was simply to be finished with school. Truth be told, it was difficult to think past graduation. After all, I had been in school my entire life. However, it seemed simple. I would graduate from high school and get a job. In fact, my parents, who had given up all hope of me going to college at that point, informed me that after high school, they would start charging me rent for my bedroom. All I cared about was being finished with high school. Then, I would be happy.

Shortly after my junior year ended, I had a moment of truth. First, I must preface this by admitting that I have always had a huge, sweet tooth, especially for candy. I was browsing through Walmart's candy section one day. With the little money that I had, I was trying to decide between gum drops, jellybeans, and after dinner mints. Then it happened. To my right, I saw an elderly gentleman, who worked at Walmart. He was slumped over and was stacking the candy neatly on the shelf. I suddenly wondered: Was this my future without further education? Was I destined to live in the same town working a minimum wage job? Yes, high school would end, but was this my future when it ended?

The next few months, which bled into my senior year, brought on some soul searching. I knew to achieve a better quality of life that I needed more education, and this meant achieving a college degree. But how? How was I going to succeed in college-level classes if I had barely passed high school classes that were considered below average? More importantly, I did not know how to be a good student. I did not know how to study or prepare for exams. At the nexus of all my concerns was mathematics. I knew that to complete a college degree, I would somehow have to pass a college-level math class. But again, how? I knew my math skills were marginal at best. I was all too familiar with sitting in a math class and staring at the blackboard (or whiteboard) feeling completely lost, just hoping the teacher would not call on me to answer a question. I was also familiar with looking hopelessly at an exam and just knowing I would fail before answering any questions. I had made many imprudent decisions in my academic career, but I was smart enough to know that I was not "college material."

Let me be clear that a liberal arts college degree is not for everyone. Some people have special, and imperative skills in trades such as plumbing, auto mechanics, carpentry, or as a beautician. Let me also be clear that I did not have (and still do not have) any skills in any of these areas. In fact, I recall a young man in my high school winning an award for rebuilding a car engine. I would not even recognize a car engine unless it was clearly marked. Therefore, I needed to go to college.

A first step for me was deciding to begin my college career at a community college. Community colleges were at least open admissions, which meant my shoddy academic record would not negatively impact me. However, my other concerns lingered. As this book progresses, I will share how attending college allowed me to overcome my math obstacle and basically transformed me from an academic failure to a success. That is some of my story. Next, we will examine common reasons students enter college despising and fearing math.

Did You Have a Math Bully?

A student's math anxiety or hatred of math tends to stem from prior experiences. Quite possibly this student encountered a "math bully" or even several math bullies. There does not appear to be a matter-of-fact definition for a "math bully." Therefore, I am defining a math bully as someone who makes a person feel inadequate or insecure about his or her math abilities by using intimidating or humiliating practices. A math bully may be a former teacher, or several teachers, classmates, or even parents. Below are some examples for students, in their own words, regarding experiences with math bullies.

> *I really tried to understand math, but when we got to fractions and canceling and finding the least common denominator, it was just so hard for me to understand, but my teacher would just yell at me. When she asked me questions, and I didn't know, she would just yell at me louder. I remember getting so scared before math class every day.*

Daniele

> *Early on, I just gave up, and most of the time I would just write any old answer down in my homework to look like I did my homework. So, this one time in grade school, we were going over homework, and she [the teacher] called on me. I can't believe I remember this, but I wrote 5 down for the answer, and yeah, it was a random guess. I sheepishly said, "I think I did this wrong." She looked at my answer and tore into me for what seemed like forever. I just remember her saying, "How stupid of you! How stupid of you! You're stupid!"*

Otis

> *My teacher would always have good students tutor the bad students in math, and I was always one of the bad students. He [the teacher] would call the names of*

the tutors and who they were paired with. When it came to me, he would always say I was in critical condition, because I couldn't get the math. I hated math, and I always felt like crying in math class.

Mike

I got laughed at a lot in math class. We used to go around the room, and the teacher would call on us. By the time it was my turn, the kids were all snickering and laughing because they knew my answer was going to be stupid.

Larry

I love my dad, and he always wanted me to succeed, but he made learning math really hard. He would sit with me every night in elementary school, because I had so many problems learning math, but instead of encouraging me, he would always yell at me. He had no patience. It seemed like as a I got older, his patience got worse. He almost got angry at me before I started my math homework. He used to yell at me to concentrate and follow the directions and told me my answers were stupid.

Bob

Math class in high school became such a total nightmare. My answers in class, which were always way off, became a joke for the class. The other kids even started posting my answers on Facebook and Instagram. I wound up dropping the class.

Derek

This one time this guy [the teacher] made me stand up from my seat, and he said, "How can you be so beautiful but be so stupid?

Audrey

Loss of Control and the Fear of Feeling Stupid

Did you know that a major source of anxiety is the feeling of losing control of a situation or control in our lives (Peterson, 2015)? As a major control freak, I can relate to this. When I get anxious, I can often identify the root cause of my fear as the loss of control, but what does this have to do with math? For many students, struggling to keep up in math class can make students feel as though their academic careers, or even their lives, are spinning out of control.

I just knew when I fell behind or even if I was absent from school, I would be in trouble with math. I knew I would be lost in class, and I hated that feeling. It felt like everyone else was getting it, and I just had to nod my head like I was getting. I was so afraid the teacher would call on me, and everyone knew I was lost.

Bob

Math is so totally different than other subjects. In English and history, you can usually look up the answers in your textbook. You know, you have a homework assignment with questions and the answer is somewhere in the reading. But that's not the case for math. You need to understand how to do a problem. Like I always struggled with factoring in algebra. No textbook was going to help me. If I didn't get it, I didn't get it.

                                                                                    Joanne

I was afraid to go back to school [college], because I knew I had to pass math. And I could just see myself in a math class trying to keep up but just not being able to and knowing I was going to fail because I couldn't keep up. That alone would give me a panic attack.

                                                                                    Daniele

I knew I needed to pass math, and I just felt so lost in class. I was trying to understand what was going on, but I just couldn't. Failing math was going to screw my life up because I would have to go to summer school and maybe not graduate. I just felt like my life was out of control.

                                                                                    Derek

All of us have felt stupid at times. Whether we are in elementary school, college, or even graduate school, the reality is no one wants to feel foolish in any situation.

I was never good at math. I always felt like such a slow idiot when I was in my math classes in grade school and high school. So, I just didn't want to go to school [college] and feel so stupid taking math again. I mean feeling like an idiot makes you cry, and it makes you feel like you are just bad at everything. I already struggled with anxiety and depression, so who needs that?

                                                                                    George

Math Feels Like a Poor Fit

Many students simply feel math is a poor fit for their personalities and learning interests and, consequently, feel lost.

I've just never been a numbers person. I like reading, and I want to become an English teacher someday, but I didn't want to go to school because of math. I just felt so out of place in math classes.

                                                                                    Joanne

I'm a dancer, so dance and music are my thing. I love dance classes. I can't explain it, but in my dance classes, I feel like I'm with kindred spirits. All of us share this common love. With math, I'm just a stranger in a strange land.

Alicia

A math class for me was like going to a party or gathering where I just didn't fit in. I just felt this immediate disconnect to the teacher and to other students who "got it" because they were all speaking the same language, you know the math language.

Jim

I will add to this by saying that I can understand this relating to another discipline. Many students have "math anxiety." Well, I have "philosophy anxiety." I can remember taking a graduate class in ethics and discussing concepts such as "the self," "the other," "I–Thou relationships" and all these various "states of being." I did not understand those concepts (and still do not), and I consequently felt very out of place. I am a literal and matter-of-fact thinker, and I simply cannot relate to those abstract concepts. I used to drive my professor crazy, as he would say, "Brian, you're trying to take these abstract concepts and put them into these square boxes." My response was, "Yes, that's exactly what I am trying to do." Most of all, I remember how other classmates became enamored and engaged in various philosophical discussions, and I felt like a complete fish out of water.

Math is Viewed as Authoritative

A major reason people fear and hate math is that math comes across as very authoritative in that there are several rules. Of course, that would strike fear or cause feelings of disdain. We can all remember that imposing person who told us to "follow the rules and that those who do not follow the rules will be punished."

Yes, math is full of formulas and step-by-step procedures. These are known as algorithms. I will provide some examples later in this book, but algorithms save us time and frustration. Think of an algorithm as a shortcut. If you need to drive to a specific destination, would you rather take the highway and other main roads, which would take 10 or 15 minutes, or would you rather take a bunch of backroads that would take 45 or 50 minutes? We will discuss why it is important to understand what we are doing when solving a math problem and to think everything through; however, it is also imperative to utilize a series of steps.

There are also math rules that students are taught that make no sense, as early as elementary school; yet students are forced to memorize these rules. We will address some of these rules in each section entitled, "Why in the World?"

Ambiguity Should be a Four-Letter Word

If I were to ever develop a phrase that was to be my legacy, I would request the following, "While ambiguity is not a four-letter word, ambiguity should be a four-letter word." It is important to understand that struggling is a part of learning. New concepts can certainly seem unclear, and they can take time to master. However, students deserve matter-of-fact and thorough instruction in mathematics, and too much ambiguity leads to frustration and anxiety.

> *It's probably weird I remember this, but I'll tell you why. This one time in eighth grade, we were doing graphs. I don't remember the exact equation, but let's say it was something like 3x+4y=12. We needed to graph the equation, and my teacher just substituted 0 for x and solved for y. I didn't get where she came up with the zero. It was literally the only time I ever raised my hand to ask a question. I asked, "Where did you get the 0?" She said, "I made it up," and she moved on. I was thinking, "So that's math? We just make up numbers?" All these years; that just stood out.*
>
> Mike

> *I hated absolute value. What frustrated me was no one could give me a clear answer of what the heck absolute values was! Just tell me in plain English what it means!*
>
> Peter

> *I could never figure out how to use those stupid greater than or less [< or >] than signs. Teachers always yelled me because I got them wrong. But I always felt like if someone could just explain them to me clearly, I could have gotten it.*
>
> Bob

> *Word problems never made sense to me. I just never knew where to even start. I would read the word problem and just be lost.*
>
> Alicia

Let me pose a question. In real life, doesn't ambiguity cause frustration and anxiety? If you are concerned about a situation and have questions, and you are not getting clear and concise answers, this ambiguity can produce anxiety. What if you are anxiously awaiting the results of a serious medical procedure and you receive the results, but you are not understanding the results or you are not receiving a matter-of-fact answer from your doctor? Wouldn't that frustrate you and add to your anxiety? Additionally, ambiguity may drive a math bully. Bob mentioned how his father was a math bully. However, when parents struggle with math and do not understand how to help their children, this can lead to anxiety and frustration.

Math is Required for Every Degree. Who Did This to Us?

Here is the reality. A college-level math class is required for just about every degree. For students, and there are many, who complain about this, I tell them that they were born too late. Math has been taught in American Higher Education since its inception in the 17th century. In these inaugural years of American Higher Education, the curriculum focused on liberal arts classes such as math, Greek, Latin, ancient history, and ethics. There was an emphasis on drill and practice; however, the late 19th century saw the beginning of a progressive movement in education. Progressives stressed that school should focus on socialization, critical thinking, and problem-solving skills. Consequently, they did not see the value that "drill and practice" mathematics added to education. Therefore, from the late 19th century through the first part of the 20th century, it was common for mathematics not to be required for a college degree (Grouws, 1992; Tucker, 2013). Certainly, students completing STEM (Science, Technology, Engineering, and Mathematics) degrees needed to complete math classes. However, non-STEM students may not have had to complete a math class.

The nation's view toward mathematics changed in the middle of the 20th century. This was due to the United States' victory in World War II with the construction of the atomic bomb. Additionally, the Soviet Union's launch of Sputnik, which started the race of the space age, enhanced the nation's view toward math. The country began to view math as an integral part of national defense and national pride (Barlage, 1982; Tucker, 2013). Consequently, math became a required course in college degree programs. This is why college algebra initially became the main course that students had to pass to complete a college degree. College algebra typically contains the content required to complete a calculus course, and calculus contains the content that the nation valued after World War II and Sputnik (Taylor, 2017).

Just because the nation or higher education sees value in learning mathematics does not mean students see such value.

> *I want to be a dancer. I just didn't see the value in learning algebra. How was struggling through algebra or any kind of math going to help me in my dancing career?*
>
> Alicia

*It's just so hard to really try hard in a subject when you don't really see how it
can help you, other than you need to pass it.*

　　　　　　　　　　　　　　　　　　　　　　　　　　　　　　Bob

The reality is that students need to see value in their coursework, and failure
to do so can negatively impact their results. I can relate to this as well. When
I transferred from my community college to my university, I had to enroll
in a two-credit transfer English class, because I was two credits short in my
freshmen-level English requirement. I was resentful that I, as a college junior,
had to waste my time with such a class when I was trying to focus on my
degree requirements. I decided that the class was a waste of time even before
the first class. I passed the class, but I did not do as well as I could have.

Returning to School After Some Time Away?

It is unfortunate that we still must identify students who are ages 18–22 as
"traditional college students." The reality is that many adults who are of
"non-traditional" age attend both community colleges and four-year schools.
As of the fall of 2021, over 6 million college students, in the United States,
were over the age of 25 (National Center for Educational Statistics, 2021).
Nonetheless, returning to school can be very intimidating for adults who
have been out of school for some time.

*I was having anxiety attacks the entire week before school started. The morning
of my first class I couldn't stop throwing up. It was so many things. I couldn't
believe I was going back to school after 30 years. I was going to sit in a classroom
with all those kids. On top of that, I had to take math! Everyone was going to
think I was just some stupid old guy.*

　　　　　　　　　　　　　　　　　　　　　　　　Otis (48 years old)

*I wanted to go back to school. It was a dream of mine to get a college degree, but it
seemed so impossible. When I would picture myself in a classroom with all these
young kids, it started feeling more like a nightmare.*

　　　　　　　　　　　　　　　　　　　　　　　George (39 years old)

The list of concerns for returning students can be long and winding. Students
worry about fitting in to what they perceive as a young population. As if
attempting math is not scary enough, imagine attempting the subject years
after being in school with ample time to dwell on miserable math experiences.
Additionally, these students have concerns regarding technology. Technology

advances rapidly and returning students may question whether they can keep up with their studies and technology. In summation, if you are returning to college after some time, you are not alone, and you can do this.

Are Men "Better at Math" Than Women?

Overall, math anxiety is higher in women than in men. In fact, some females begin to experience math anxiety in early elementary school (Van Mier et al., 2019). The unfortunate reality is that many women feel inferior to men in the discipline of math. In fact, women are still underrepresented in professional fields such as math, computer science, and engineering (National Science Board, 2018).

In our present society, there are women who are mathematicians, engineers, chemists, and math educators, while there are men who suffer from intense math anxiety. Personally, I have known many women with superior math skills. In fact, when I tutored math in college, I worked alongside a woman who I still consider to be the most talented mathematician I have ever met. In fact, just to entertain ourselves on slower days, in the tutoring center, we would open a calculus book, turn to the very back, where the most arduous problems exist, and ask her to complete those problems, which she did with such ease. No proof exists that men are better than women in math. In fact, there are extensive biological and cognitive studies that show that men do not have any advantages in math over women. Additionally, there are no significant studies that show that men outperform women in math. So, why do women have less confidence in math?

When conducting research on the gender issue in math, it seems like many people have varying opinions. However, in many cases girls and women have been treated as if they were inferior in math or even told that they were not as good at math as their male peers by teachers and others. Consequently, many females have come to believe this. In some cases, the treatment is subtle. For example, female students may be called on in class less than their male peers to answer questions (Lavy & Sand, 2015). In other cases, some elementary school teachers believe that girls must work harder than boys in math and, by default, need extra help. Additionally, many female teachers at the early elementary school level struggle with math anxiety, and this, in turn, impacts female students and their feelings toward math (Beilock et al., 2010). By high school, many young women are subtly driven away from STEM fields. This is especially the case for Black and Latino young women, Again, why is this? There is no simple answer, but the higher education timeline may shed some light on this.

While American higher education has existed since 1636, the first generation of women did not enter higher education till the late 1800s. There are reports of various women attending colleges and universities before then, but they were few and far between. When women began attending institutions of higher education, they largely populated disciplines such as nursing, education, and social work. The nation largely viewed these fields as "women's fields" as opposed to math, science, and engineering, which were overwhelmingly populated by men. This trend continued throughout most of the 1900s. It wasn't till the latter part of the 1900s that traditionally male-dominated fields began to see an influx of females. When looking at American higher education through this wholescale lens, there have only been a few decades where women have significantly enrolled in the math, science, and engineering fields compared to a few hundred years of male domination. Hopefully, one day, the misconception that STEM fields are "male fields" will be a thing of the past. Unfortunately, this has been a slow process.

Addressing Racial Inequalities in Math

Women are not the only ones who may feel inferior in math. Unfortunately, studies have shown that Blacks and Latinos struggle in math as well. At the college level, Latinos and Blacks are more likely to be placed into developmental math classes and have lower success rates in math compared to White students (Bahr, 2010). Like the gender issue, there is no proof that White students are any better at math than Latino or Black students. So, what gives?

The struggles for Latino and Black students can be traced back to K-12. Black and Latino students are unlikely to attend schools that offer advanced math and science programs. Additionally, a high population of Black and Latino students attend inner-city schools and struggle with poverty issues such as food insecurity as well as safety issues. When students struggle with these issues, it makes learning secondary and difficult. A major goal in education must be to help minority students fill in mathematical gaps.

Like women, Black and Latino students may be led to believe that they are inferior in math. They are less likely to be called on in class and may be eased away from STEM fields. Fortunately, organizations such as the National Science Foundation have implemented programs such as the Louis Stokes Alliances for Minority Participation (LSAMP) and Women in STEM (WiSTEM). LSAMP and WiSTEM serve as outreach programs to help more minority individuals and women enter the STEM fields.

Overcoming Defeatist Attitudes

Many students hinder their math experience with pre-existing defeatist attitudes. Here are some examples:

> *I could never handle fractions. I just knew if I encountered fractions, I wouldn't get them, and I would get the whole problem wrong. Whenever I was working on a problem, and fractions came up, I just got frustrated and gave up.*
>
> George

> *I always had a problem with long math problems with multiple steps, like those three-by-three system of equations. I just knew if it was a long problem I would get lost somewhere in the middle. I would even just tell my teachers at the beginning of the term that I just couldn't get long problems. I expected them to just cut me a break.*
>
> Bob

> *Word problems always got me. I just looked at a word problem and would give up. It was just years of failing at word problems. I once went off on one of my teachers that people invented word problems just to confuse you.*
>
> Tatiana

Does any of this sound familiar? For many students these pre-existing pessimistic attitudes often lead to failure. In the mental health field this is known as a self-fulfilling prophecy. This is when an expectation or belief can influence a person's behavior and a situation's outcome. Changing these pre-existing attitudes is not easy and can take time. However, understanding that you may be hindering your own success with an incoming negative attitude is the first step.

The Development and Treatment of Math Anxiety is a Process

While many students can recall various terrible memories of math, it is likely that a series of events not just one event led to their math anxiety. Consequently, eliminating math anxiety is not as easy as flipping a light switch. As we will examine in the following chapters, it takes careful planning prior to entering school, smart choices, preparation, discipline, and learning from mistakes. Applying all of these can ease math anxiety and gear you toward success!

Why in the World? Part 1

As mentioned earlier, in each of the first eight chapters, we will examine a mathematical topic from arithmetic or basic algebra that, on the surface, makes no sense. "Why in the World" will feature mathematical topics that seem like useless and made-up rules that students must memorize.

Why in the World Does Multiplying a Number by Zero Give Us Zero?

For example, $2 \times 0 = 0$, $5 \times 0 = 0$ and so forth. Students have been exposed to basic multiplication, and these problems, since the second or third grade; yet, it is a rule that baffles many students.

Why this makes no sense: Consider 2×3, which can also be written as $2 \cdot 3$ or $(2)(3)$. Students assume this means we take the number 2, three times, which would be $2 + 2 + 2$, and that is 6. So, how can we start with 2, take it zero times and wind up with zero? Also, multiplication is supposed to make everything larger. Even in subtraction, where numbers get smaller, if we subtract $2 - 0$, we still have 2!

Let us try to understand: Multiplication is about grouping. The problem 2×3 is stating that there are two groups of three, meaning two groups with three cookies in each group giving a total of six cookies.

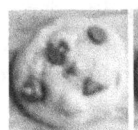   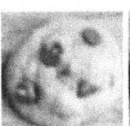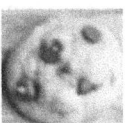

Below are two more examples:

2×2 (2 groups of 2)

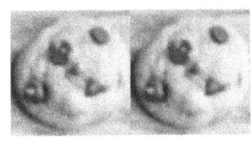 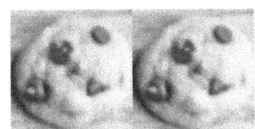 = 4 (total cookies)

2×1 (2 groups of 1)

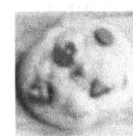 = 2 (total cookies)

Now, we get to 2×0. This means two groups of zero, meaning two groups containing zero items in each. That would look like this.

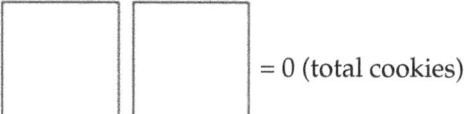 = 0 (total cookies)

Therefore, two groups of nothing totals nothing.

Activity

Before enrolling in a math class, it is important that you reflect on your prior math experiences and more importantly reflect on what led you to dislike and fear mathematics. Your math autobiography (Arem, 2003) should include answers to the following questions:

- Did you have a math bully or multiple math bullies? How did these people make you fear and dislike math? What would you say to them now?
- Has math made you feel inferior to your peers? Or has math made you feel like a poor student? What do you feel that it takes to be successful in math and do you feel like you are capable of that?
- What are your favorite subjects? Where do you feel that you best fit in?
- Do you feel that learning math has value for you as a student? If not, what would it take for a math class to contain value for you?
- Can you identify any defeatist attitudes that you carry into your math classes? If so, are your defeatist attitudes generic or centered around specific areas in math?
- In this chapter, I discussed some of my struggles and how those struggles served as a barrier to my learning math. Have you had any external struggles that have led to your struggles in math? What were they?
- Develop a math anxiety timeline for yourself. When did your math anxiety start? How did it progress?

References

Arem, C. A. (2003). *Conquering Math Anxiety* (2nd ed.). Brooks/Cole: Grove, CA.

Bahr, P. R. (2010). Preparing the underprepared: An analysis of racial disparities in postsecondary mathematics remediation. *Journal of Higher Education, 81*(2), 209–237.

Barlage, E. (1982). The new math: A historical account of the reform of mathematics in the United States of America. (ERIC Document Reproduction Services No. ED 224703).

Beilock, S. L., Gunderson, E. A., Ramirez, G., & Levine, S. C. (2010). Female teachers' math anxiety affects girls' math achievement. *PNAS Proceedings of the National Academy of Sciences of the United States of America, 107*(5), 1860–1863. https://doi.org/10.1073/pnas.0910967107

Blazer, C. (2011). *Strategies for reducing math anxiety* (Report No. 1102). InformationCapsule Research Services: Miami Dade Public School. https://files.eric.ed.gov/fulltext/ED536509.pdf

Grouws, D. A. (Ed.). (1992). *Handbook of research on mathematics teaching and learning: A project of the National Council of Teachers of Mathematics.* Macmillan Publishing.

Lavy, V., & Sand, E. (2015). *On the origins of gender human capital gaps: Short and long term consequences of teachers' stereotypical bias* (Paper 20909). National Bureau of Economic Research. www.nber.org/papers/w20909

National Center for Educational Statistics. (2021). https://nces.ed.gov/ipeds/TrendGenerator/app/answer/2/8

Peterson, T. J. (2015). *Anxiety and your sense of control.* Healthy Place. https://aws.healthyplace.com/blogs/anxiety-schmanxiety/2015/12/anxiety-and-a-sense-of-control

Science and Engineering Indicators. (2018). *Early gender gaps in mathematics and teachers' perceptions.* National Science Board. www.nsf.gov/statistics/2018/nsb20181/assets/481/early-gender-gaps-in-mathematics-and-teachers-perceptions.pdf

Taylor, S. (2017). *The evolution of college algebra: Competencies and themes of a quantitative reasoning course at the university of Kentucky* [Doctoral dissertation, Western Kentucky University].

Tucker, A. (2013). The history of the undergraduate program in mathematics in the United States. *The American Mathematical Monthly, 120*(8), 1–21. https://doi.org/10.4169/amer.math.monthly.120.08.689

Van Mier, H. I., Schleepen, T. M. J., & Van den Berg, F. C. G. (2019). Gender differences regarding the impact of math anxiety on arithmetic performance in second and fourth graders. *Frontiers in Psychology, 9,* Article 2690. https://doi.org/10.3389/fpsyg.2018.02690

2

I'm So Confused... Navigating the Math System

In the past, enrolling in a math class was very simple. After being admitted to college, the student signed up for and completed a math placement exam. The results on the placement exam yielded a raw score, and that raw score dictated which math class the student started. If this was a developmental (remedial) math class, the student generally had to complete several classes to complete either college algebra or introduction to statistics. Of course, the above scenario still exists. However, students enrolling in a college math class generally have more options in the present day. In fact, there are alternative math pathways that are tailored to the needs of students. Still, these options may be confusing for students. In this chapter, we will focus on possible options for enrolling in math classes, so that you can make better choices.

First Things First

Your first step toward a college education is the application process. Of course, there are preliminary steps that you can take as well. Most colleges offer tours to prospective students. You can tour the campus and ask questions as well. Many colleges also host enrollment fairs or varying information sessions prior to enrollment. This is an opportunity to not only view the types of degrees offered but talk with faculty and staff at the college as well. Applications are generally found online on the college's website. You will likely need to create an account and provide various types of information regarding your academic background. You may also need to submit transcripts from your previous schooling.

As for acceptance requirements, community colleges are open admission. Some community colleges may require students to possess a GED. This stands for a General Equivalency Diploma or the completion of General Education

DOI: 10.1201/9781003614142-2

Development Tests. However, other community colleges may have more flexible admissions policies and may, consequently, not require a GED. No one is judging you based on your background. This is all a formality. It does not matter what you did. What matters is what you do going forward.

Four-year colleges likely have acceptance requirements. In the past, most four-year schools required students to have achieved a minimum score on the SAT exams. This has changed, as many colleges no longer require minimum SAT scores. However, such colleges likely require a minimum high school grade point average (GPA). Of course, there are universities that require minimum SAT scores. The bottom line is that the admissions requirements for four-year colleges vary greatly. That is why it is imperative to research this on the college's website or to visit with an admissions counselor prior to applying.

Choosing a School

Community Colleges

There is generally only one community college per county, and it generally costs more to attend a community college that is out of county. Additionally, most community colleges offer all general education classes that students must take in their freshman and sophomore years, in subjects such as mathematics, English, and social sciences. Therefore, many students choose a community college based on proximity. However, there are some factors to take into consideration. First off, is the community college accredited? All community colleges report to a specific accrediting agency. Accreditation is the recognition that the community college maintains a certain level of educational standards. A community college that is not accredited does not hold a solid reputation with other colleges and within the community. This could present a problem for students when they transfer to a four-year school or look for work. Second, when choosing a community college, consider your career goals. If you are looking at a two-year degree in business or nursing, for example, examine those programs. Are they reputable?

Four-Year Colleges

Students often choose four-year colleges based on their degree goals. For example, a student looking to major in business may search for a college with a solid business program. This is the same for degrees such as education, allied health, or engineering. This may require students to attend a college in a different part of the state or in a different state altogether. However, as the 21st century has progressed, more college students commute, as opposed to dorm. In fact, over 80% of all college and university students commute to school (Jacoby, 2020). Consequently, many students may even choose four-year colleges based on proximity.

Online Institutions

In the present day, students may choose between several online institutions such as the University of Phoenix, Western Governors University, or Southern New Hampshire University. Such schools will advertise their flexible schedules through online classes. However, you should not assume that online classes or degrees are easier. Later in this chapter, I explore the advantages and disadvantages, or challenges, of online classes. Additionally, there are companies or institutions that will only hire students who possess degrees from traditional colleges. Therefore, you should conduct some research by speaking with people in companies or institutions who are in your desired field (e.g., business, education, etc.) to determine if they will hire people who possess degrees from fully online institutions.

Financial Aid and Scholarships

College can be expensive, and how expensive varies among institutions. Of course, terms such as "expensive" and "cheap" can be subjective. Community colleges offer comparatively low tuition; however, this may still be expensive for students. The cost of books and other classroom materials can be costly. Students who dorm at college face housing costs as well. That is why it is imperative to research ways of paying for college.

Financial aid is available in varying degrees for students. Students can complete a FAFSA (Free Application for Federal Student Aid) form online. However, it is imperative to make wise choices when applying for financial aid. Immense student load debt has become a huge problem in America. Such debt results from students making uniformed decisions and taking on greater loans than they can eventually handle. This is why it is a good idea to meet with a financial aid advisor before applying. You will want to get an idea of how much money you should apply for and an idea of your eventual student loan payments.

Scholarships are also available for students. Scholarships may be for academics or for sports. It is imperative to conduct extensive research on what scholarships are available. A college's financial aid office may have information regarding scholarships. For high school students, guidance counselors can typically assist with this information.

The Placement Test

Most colleges require incoming students to complete a placement test as a first step. This is to help the student start in a class, such as math or English, in which the content parallels the student's incoming skill level. Many colleges utilize standard placements that are adaptive. More specifically, subsequent questions are chosen based on a student's previous response. If a student continues to answer questions correctly, he or she may receive more difficult questions. There are, however, schools that utilize their own placement exam. There are generally cut scores or intervals that dictate placement.

Throughout the 2010s, many colleges started using multiple measures for incoming students. This means that in addition to the placement test scores, students may be placed in a class based on factors such as high school grade point average (GPA) or scores on the ACT or SAT (if applicable) exams. The use of multiple measures has been growing, as many administrators and educators have expressed concern that placement exams, alone, do not place students with complete accuracy.

What is Developmental Math, and Will I Have to Take It?

The first part of the above question is simple. Developmental math courses consist of math content that is below college level. Content in such classes is considered redundant in that students were supposed to have mastered it in high school or even elementary school.

The second part of the question is a bit more complex. Stand-alone developmental math classes have existed and served underprepared students since the mid-1800s. With the increased use of placement testing in the latter part of the 20th century, students were and, to some degrees, still are, placed into a developmental math class if their score is too low. In the past, some students needed to complete multiple developmental mathematics classes. However, in recent years, colleges have worked to shorten math pathways, especially for non-STEM (Science, Technology, Engineering, and Mathematics) students. More specifically, students who are not entering a math-related field may only need to take one or two developmental math classes before attempting a college-level math class. I will expand on this later, but it is also possible that students can complete their developmental coursework at the same time as their college-level math classes.

What is "College-Level Math?"

A math class is considered "college-level" by the state. These math classes generally transfer to and among colleges and universities. Developmental math classes, in contrast, generally do not transfer or satisfy a student's degree requirements. Some colleges do not even award credit for these classes. Traditionally, college algebra and introduction to statistics have been the main college-level classes that have satisfied degree requirements for non-STEM (Science, Technology, Engineering, and Mathematics) majors.

Oftentimes, community colleges offer math classes that satisfy two-year associate degrees. Generally, these are business math classes that satisfy an associate of business degree and allied health math classes that satisfy various allied health associate degrees. While these classes satisfy an associate degree, they do not transfer to a four-year college or university. Consequently, students need to complete a course such as college algebra or introduction to statistics to complete a four-year degree.

The Traditional Pathway

As mentioned earlier, traditionally, when a student enrolled in college, there was basically a one-size-fits-all pathway. In most cases, the students, who were non-STEM (Science, Technology, Engineering, and Mathematics) majors needed to complete college algebra or possibly introduction to statistics to complete their math. However, the pathway could be long and winding. Oftentimes, students placed into developmental math, and in many cases, this was low-level algebra or arithmetic. This meant these students needed to complete up to five or six math classes simply to arrive at a college-level math class.

As the 2000s progressed, college math began to change. Researchers discovered two problems. First, an overwhelmingly high number of students placed into developmental math and too few students ever completed their college-level math requirement, much less got out of developmental math. Second, college algebra is not a fit for all students. For example, college algebra typically contains content such as various types of functions and conic sections, which prepares students for calculus. However, how does this content help students who do not intend to continue with math, such as nursing, sociology, dance, or English majors?

Alternative Math Pathways

Students who are non-STEM majors have more choices now. Below are alternative math pathways and the explanation of corequisite classes.

Quantitative Reasoning (or Math for the Non-Math Major)

Quantitative reasoning (QR) is a college-level math class designed for the non-math major. QR consists of real-world applications focusing on basic statistics, probability, and finance. This has been a popular course as students appreciate how this content relates to everyday life. To be successful in this course, students need a strong arithmetic background (operations with whole numbers, fractions, decimals, and percentages) and some pre-algebra (solving basic linear equations and equations with fractions). QR courses consist of many word problem applications. Therefore, being able to interpret and solve word problems is imperative as they require solid reading comprehension skills. Keep in mind different schools may have varying names for this class. Essentially, this is a math class for non-math majors.

QR is generally a terminal math class. If the course is in the state's transfer module, QR, if successfully completed, will transfer to a four-year school and could be the final math course for a non-STEM student.

Who should enroll in a QR course? Yes, a non-STEM major, but let us be more specific. QR is a course for students who will never need another math class throughout their college career. For example, this will most likely include fine arts majors, English majors, or sign linguists' majors.

There has been much debate regarding the best pedagogical practice in a QR course. Educational policy makers, administrators, and even some faculty have argued that in addition to mastering the mathematical content, students should be able to think critically. Therefore, such individuals have pushed for QR to utilize inquiry or group-based instruction. This is when students must master the content by working together in groups and problem solve with classmates. Inquiry takes more of a bottom-up approach to learning, as opposed to a traditional top-down approach where the professor guides the learning process. The inquiry method can lead to more critical thinking and can help students master the material better, as they are learning through their own discovery; however, this method can also lead to confusion and frustration for students. Several students have asserted that they would prefer to have their professors walk them through new material in math. More specifically, these students prefer the scaffolding approach where the professor breaks down and introduces the new material into manageable parts and then allows the students to practice the material before moving forward. Later in this chapter, I will provide a more specific comparison of traditional instruction to inquiry-based instruction. I will discuss the importance of advising later in this book; however, students should inquire, through advising and other sources, about a professor's teaching style. Regarding a QR course, determining whether a teacher uses inquiry, or a more traditional practice, would be imperative information.

Introduction to Statistics

Colleges and universities have been offering introductory statistics courses since the 1700s. However, students used to have to completely test out of developmental math or pass the highest level of developmental math to enroll in a statistics class. Gradually, more colleges are allowing students to enroll in introduction to statistics with simply a background in pre-algebra. However, as I will discuss in Chapter 3, introduction to statistics is a fast-paced and deep class.

Are introduction to statistics courses for non-STEM majors as well? Yes, but it is not that simple. Students who major in social or behavioral sciences such as psychology, sociology, or economics often need a background in statistics later in their education endeavors. More specifically, more advanced courses in these fields require students to conduct and interpret statistical studies. Therefore, in summation, students entering fields in the behavioral or social sciences should consider introduction to statistics for their college-level math requirement, whereas students entering fields in fine arts may find QR as a better fit.

Teacher Preparatory Classes

Are you interested in becoming an elementary school teacher? If so, you will likely need to complete one or two math teacher preparatory courses. These are classes specifically designed for future elementary or middle school teachers. In some cases, these teacher preparatory classes satisfy the math requirement for a degree. However, some schools require that students complete an additional math course such as statistics or college algebra.

What content is covered in these teacher preparatory courses? Again, that is another complex question. The content covered may vary from state to state or even school to school. However, in recent years there has been more of a push from certain states to promote more of an inquiry-based format regarding instructions. Students work together to solve abstract problems and utilize manipulatives or mathematical games.

What are the mathematical prerequisites for these teacher preparatory classes? Students should have a solid arithmetic background in topics such as whole numbers, fractions, decimals, geometry, operations with percentages, and word problem applications, as future teachers should know these topics for their future careers. Additionally, some teacher preparatory classes utilize algebra concepts such as linear and quadratic equations, evaluating expressions, and system of equations. Students may also be required to write research papers and construct their own lessons. Oftentimes, students may need to complete fieldwork by assisting at an elementary school to gain experience.

What are Corequisites?

Corequisites involve the pairing of developmental coursework with college-level coursework. This is also known as "just in time remediation." Corequisites are typically used with introductory college-level courses such as quantitative reasoning, introduction to statistics, or even college algebra. In many cases, students register for two different classes: the college-level class and the corequisite class (sometimes referred to as the "booster class"). In the booster class, students focus on the developmental content they will need for the college-level class. For example, a QR booster course may focus on arithmetic or pre-algebra concepts. A college algebra booster course may focus on elementary or intermediate algebra concepts.

Corequisites are relatively new to college mathematics. They serve two purposes: to allow students to complete their math requirements at a quicker pace and to allow students to apply new knowledge from their developmental coursework almost immediately to their college-level class without waiting an entire semester to do so. As for scheduling, schools have varying ways of arranging these booster classes. Some colleges schedule the booster classes immediately before the college-level classes. More specifically, students will study the remedial content in the booster course, which could last an hour or two, and then take their college-level classes afterwards. Other schools offer their booster classes on different days than the college-level classes. For example, the booster class may run on Mondays and Wednesdays, while the college-level class runs on Tuesdays and Thursdays.

The corequisite model has made a positive impact in college math. Students are moving through their math requirements at a quicker pace and appreciate the opportunity to do so. Students also value being able to learn and apply remedial material quickly. However, students must also understand that taking a booster course alongside a college-level course is a lot of work. Not only does it require extra time, but the corequisite model requires students to learn remedial content and quickly apply it to the college-level content. Fatigue, from extra work, can be a factor, but the pacing is also very fast.

Depending on the school, or possibly the state, it is possible that the college may not even offer a stand-alone developmental class. Consequently, students will enroll in a college-level math class, by default, with a booster course, and this booster course would attempt to concurrently prepare students for the college-level class. This certainly saves students time, but many students struggle with trying to learn all the needed remedial material while taking a college-level math class.

The STEM Pathway

Are you considering a future career in engineering? This could include environmental engineering, computer-assisted design, or aviation. How about a career in physics or chemistry? Has biology interested you? How about forensic science? The reality is that to complete a degree in any of the aforementioned fields, students need to at least complete a calculus course. To complete calculus, students need to successfully complete college algebra, pre-calculus, and trigonometry. Consequently, math pathways within these STEM fields are longer. It depends on the school, but students who test into a course lower than college algebra may need to complete multiple developmental algebra classes to become prepared.

Undecided?

Many students enter college simply undecided about their future, and that is all right. Several students view college as a vehicle to a higher quality of life. Some students may have held low-paying jobs in the past and see a college education leading to a more stable lifestyle. Others have seen family members or friends prosper from a college education. Some students may have multiple career interests, and quite frankly one of the purposes of a liberal arts education is to give students a taste of various fields such as psychology, sociology, English, or economics. Of course, this can muddy the math pathway. If you are undecided about a career path but are certain that you do not wish to enter a STEM field, you should consider taking introduction to statistics over QR. As mentioned above, statistics can cover more fields. College algebra should only be for students entering the STEM fields; however, it is possible that individual program directors of non-STEM fields may still require college algebra.

Modality Types

When most students picture a math class in a college setting, they generally envision a lecture-based class. More specifically, students may imagine a class size of 100 with a professor simply lecturing to the students with no interaction between the two. However, as of 2021, according to ThinkImpact, the average size of a community college math class ranges from 25 to 35 students. Of course, some four-year colleges may contain class sizes up to 200 or 300.

However, there are varying ways in which math classes are offered. In this section, we will examine the different types of modalities in which colleges offer math classes. More importantly, we will inspect the advantages and disadvantages of each type of modality, so that you can make informed choices.

Traditional Classes

Traditional math classes are still referred to by many colleges as "lecture-based classes." In its purist form, lecture typically means that the instructor speaks, and students listen with no interaction between the two. However, throughout recent decades, college faculty have been working to improve student engagement, and this includes more interaction between the professor and the students. This is generally due to findings that straight lecture does not lead to productive learning. Therefore, it is very likely that an instructor will try to interact with and engage the class. However, a traditional class typically indicates that students attend class with guided instruction from their professor. The instructor will cover specific sections or topics in each class, and students will be given the opportunity to take notes and ask questions. Generally, students are assigned homework based on the content covered in each class.

Advantages:

- This can be a good fit for students who struggle with math anxiety and desire matter-of-fact and thorough instruction from their professors.
- A "traditional math class" is what many students are used to from their previous schooling, and this may be less intimidating (than other models) for students.

Disadvantages:

- With the instructor needing to cover specific content each class, it is possible for students to fall behind. This can be the case if students miss class.
- Some students may feel intimidated to ask questions or even be called on due to their math anxiety or low confidence in math.
- If the instructor utilizes mostly lecture, there may not be as much "hands on" where students get to practice lots of problems in class, and consequently, students may not learn as much in class.

The Emporium Model

The emporium model refers to classes held in a computer lab that is self-paced. This model has been around since the 1970s; however, it gained steam in the 2000s, as colleges sought ways to reduce the amount of time students spent in developmental math.

In an emporium model, students work at their own pace, within certain time limits, on their math. Generally, colleges use interactive math software programs such as My Lab Math, Hawkes, or ALEKS for these classes. Students will watch short video lectures to start a topic and then complete math problems on their own. Once the students have completed enough sections, they will take an exam and continue. In the emporium model, the instructor acts more as a facilitator. The instructor typically circulates throughout the lab and answers questions. As the emporium model may have larger class sizes, there may also be one or two tutors to assist the instructor. This also ensures that students' questions are answered in a timelier manner. Colleges generally use the emporium model for developmental math classes; however, some colleges use the emporium model for some college-level math classes.

Advantages:

- Students get lots of practice in class. As we will discuss, one of the best ways to minimize math anxiety and improve mathematical skills is to practice.

- The emporium model allows students to move through material that they already understand at a quicker pace. Oftentimes, sections in an emporium math class contain pre-tests, which allows students to test out of a specific section and move forward. Moreover, this allows students more time to focus on difficult areas. Additionally, the emporium model provides students with the opportunity, with hard work, to complete multiple courses in the same term.

- Students can get more individual attention from their instructors and tutors, which creates a more relaxed atmosphere. This, in turn, can alleviate a student's math anxiety.

Disadvantages:

- The emporium model can be intimidating, especially for students with math anxiety and/or older students returning to college at first. The first class, especially, may seem somewhat chaotic, as students need to register for the software, and the instructor may need to iron out all kinds of technical logistics. However, the flow of the class tends to settle down after a while.

- Students who need thorough instruction from their professors may feel frustrated, as they are expected to learn much of the material on their own.

- Again, while the emporium model is self-paced, there are deadlines. Sometimes students become too relaxed in this modality. Consequently, they fall behind and ultimately fail.

It is noteworthy that some schools offer a hybrid model as well. More specifically, a math class may be part traditional and part lab. This allows students to get thorough instruction and guided practice from their professors in a traditional classroom but also spend some time in a lab practicing math problems.

Online

Online courses consist of asynchronous learning. Generally, the professor will post various videos and instructional resources such as reading material or PowerPoint notes. It is then on the students to complete this work along with various homework assignments. Like the emporium model, online math classes typically use some sort of interactive math software. Students may be required to take their exams in some sort of proctored testing environment. If the school is too far away, students may be able to complete their exams in a library or general testing facility. Many schools allow students to complete their exams online. These colleges utilize an instrument such as Respondus or ProctorU, which, along with the webcam, can remotely proctor a student taking an exam.

Online learning has exploded since the 1990s. Nationally, from 1999 through 2005, the number of students taking at least one online class soared from 744,000 to over three million (Weisbrod et al., 2008). By the late 2010s, nearly one out of every three students in higher education was taking at least one online course (Online Learning Consortium, 2018). Clearly, online learning is a popular model; however, like all other learning modalities, there are advantages and disadvantages.

Advantages:

- For students who are working and may also have children, online classes offer flexible learning. Students can fit their schoolwork into their schedules and not have to worry about missing class.
- Like the emporium model, students can focus less on content they already know and spend more time on content in which they struggle.
- Students who have anxiety about sitting in a classroom find online learning less threatening.

Disadvantages:

- Online learning can be challenging for students with weaker math backgrounds. Oftentimes, these students become frustrated by watching videos or reading material but are not able to ask questions. More specifically, there may be slight details that confuse students when watching a video and such details may even derail their learning.

- Some students incorrectly assume that because online learning is more flexible, it is easier. Students taking an online math class will have to work just as hard as students in a traditional math class, possibly harder, as they are learning independently.

- In an online class, students are completely reliant on technology. Weak Internet connections and poor compatibility between required software and a student's laptop or iPad may be problematic, frustrating, and may impede success.

- Online learning does not work for students with poor time management skills. Students must be able to meticulously organize their time between school, work, and family.

- There is also an isolation factor. Unlike a traditional class, where they are among peers, students may feel they are completely alone in their struggles in an online class. Later in this book, we will discuss some strategies to combat isolation in an online class.

Virtual Learning

During the COVID-19 pandemic, virtual learning became a necessity. After all, the need for social distancing severely reduced and even prohibited in-person learning. However, asynchronous online learning was not enough. Students needed synchronous guided practice from their instructors and needed to learn in cohesion with their classmates, especially in mathematics. Therefore, the use of virtual instruction exploded with instructors and students using virtual tools such as Zoom and Google Classroom. The future of virtual learning is unclear in a post-pandemic world. How many virtual learning classes that colleges continue to offer in mathematics will depend on two results: the enrollment in these classes as well as the student success rates.

Advantages:

- Virtual learning offers a happy medium between asynchronous online instruction and in-person learning. Students can obtain guided practice from their instructors and learn with their classmates but also enjoy the convenience of learning from their homes.

- Several students have mentioned that they feel more comfortable asking questions in a virtual learning environment. More specifically, they feel more comfortable asking questions in a remote environment without everyone looking at them in a classroom. Students also like to employ the chat feature to ask questions. More specifically, they find this feature less intimidating than asking questions in class.

Disadvantages:

- Students can become easily disengaged during virtual classes. Some students will often multitask during a virtual class. They may tend to family obligations, work on other course material, or simply watch television or surf the web. Ultimately, this disengagement hinders student learning.
- Being removed from a physical classroom and becoming disengaged in their learning can fallaciously remove the rigor from the class. Unfortunately, this can lead students to believe the class is easier than it is, which can lead to major consequences.

Inquiry or Group-Based Instruction

I referenced inquiry-based instruction earlier in this chapter in reference to courses such as quantitative reasoning and the teacher preparatory courses. However, math faculty could utilize inquiry-based instruction in virtually any math class to varying degrees. For virtually every lesson, faculty may place students into groups and task them with problem solving with virtually no guidance or lecture. In other cases, faculty may run the class as a hybrid model in which they employ mostly traditional instruction but couple that with inquiry-based instruction. Please see Table 2.1 for a more specific comparison between lecture-based instruction and inquiry-based instruction on a topic (e.g., probability).

Advantages:

- Inquiry-based instruction promotes student engagement. Moreover, through group-based instruction, students can become part of a community and develop a connection with peers.
- Some research has shown that when students master a concept through inquiry-based instruction, they are more likely to retain that information.
- Some students benefit more from learning math through discussion, as opposed to simply being given the steps to a problem.

Disadvantages:

- Inquiry-based instruction can be difficult for students who are not used to that method. Many students are used to traditional instruction,

TABLE 2.1

Traditional Class versus Inquiry-Based Instruction

Traditional Class	Inquiry-Based Instruction
The instructor introduces probability with various definitions and provides an overview of the goals for the day and even the unit.	The instructor introduces probability to a lesser degree but does share the goals of the unit. The instructor may present a real-life application of probability to entice the students' interests.
The instructor will provide sample questions on probability and will guide the students step-by-step as to how to solve the problems.	The instructor has the students work in small groups (possibly groups of three or four). In these groups, the students are given applications of probability, and using a textbook or online material must work collaboratively to answer questions. The instructor acts as a facilitator, circulates throughout the classroom and engages in conversations with students.
After the instructor feels the students have been given enough information and understand the topic, the instructor will then allow students to work on problems involving probability either individually or collaboratively. This assessment will mirror the sample problems the instructor covered. The instructor circulates throughout the classroom and answers questions that arise.	While the instructor will assist students, he or she may not directly answer questions. In inquiry-based instruction, struggling is considered part of the learning process. Additionally, there are no limits to what students can learn. Students may cover concepts in any area of probability, as opposed to simply one part of one unit.
The class will then go over the assessment. The instructor may read off the answers to the assessment or call on students to answer questions. This will allow the instructor to move on to the next topic in probability.	The class will then go over the assessment. However, in many cases, the groups will be expected to share their findings with the class.

which is more instructor-led. Therefore, inquiry-based instruction can lead to academic shock.

- Inquiry-based instruction can cause students to get lost in the material. Inquiry-based instruction often uses real-life applications of mathematical concepts, which are important; however, when students lack basic foundational knowledge, they struggle in applying such concepts.
- There can be "people drama" associated with the group-work employed in inquiry-based instruction. Sometimes, certain groups of students simply do not mesh. Students may not get along. Some students may do all or most of the work, while others do little or nothing. This drama often impedes student learning.

What is Your Learning Style?

Before attempting a math class, it is important to determine the type of learner that you are. The reality is that you cannot control your professor's pedagogical practices. However, understanding the kind of learner you are could help you choose a professor based on reputation, and choose a modality. There are three major learning styles: auditory, visual, and kinesthetic (Arem, 2003; Dunn & Dunn, 1978). Read each question and circle either a, b, or c for a response. Notice that I mixed questions that relate to learning math but also to real life.

Learning Style Quiz

1. It is the first day of college, and I am lost. I ask someone for directions to my class. I need this person to:
    a) Verbally, tell me the directions.
    b) Write down the directions or a map on a piece of paper.
    c) Walk me to class.

2. When my teacher introduces a new topic, I learn better if my teacher:
    a) Talks as much as possible to guide me through the problem.
    b) Writes as much information on the overhead or board and even provides me with written notes.
    c) I need to work a problem out for myself as much as possible.

3. When working on a math problem:
    a) I talk as much to myself as possible.
    b) I like to read my notes over and over, and sometimes I re-write my notes.
    c) I create my own questions and answer them.

4. When I purchase a new iPhone or Android, this is most important to me:
    a) Hearing a description of the phone's features (either from the salesperson or online).
    b) The appearance and design of the phone.
    c) Trying out the phone to see if it is easy to navigate and has what I want.

5. When solving a math problem:
    a) I do not mind a lot of talking and discussion.
    b) I get easily distracted. I need a quiet place to work.
    c) I like to move around the room and take breaks.

6. When watching a math video online, I prefer:

    a) Lots of verbal explanation.
    b) Lots of visuals and written explanations.
    c) I need to be able to interact with the presenter.

7. When struggling to learn a new topic, I prefer someone to:

    a) Just keep talking to me about how to solve the problem till I understand it more.
    b) Write down sample problems and tidbits to help me.
    c) Give me real-life applications of the problem. In other words, show me how this is used in everyday life.

8. My most positive learning experiences in a math class were when:

    a) The teacher not only explained things well, but we had a lot of discussion as a class about how to solve a problem.
    b) The teacher provided notes that were clear and concise.
    c) We also learned through hands-on activities like math puzzles or games.

9. If I am struggling with a homework problem, my reaction is to:

    a) Talk myself through the process. Read the problem and talk to myself about possible solutions.
    b) Look over my notes to help me.
    c) Try to go with my intuition on how to solve the problem.

10. When trying to memorize a mathematical formula, I:

    a) Say the formula over and over to myself.
    b) Write the formula over and over.
    c) Try to come up with a song or a "catchy" way of reciting the formula, maybe use the Internet to find something.

Count the number of "a's," "b's," and "c's" that you scored. If you scored mostly "a's," you are an auditory learner. You benefit from listening to directions and guided practice. You also value interaction and dialogue with other classmates. If you scored mostly "b's," you are a visual learner. You learn best when information is presented to you in a written manner. You like to see information presented on an overhead, a blackboard or whiteboard, or a smartboard. You also learn by writing as well. If you scored mostly "c's," you are a tactile/kinesthetic learner. As a tactile/kinesthetic learner, you enjoy being physically engaged in your own learning. You learn more by practicing, but you also enjoy hands-on activities. Bringing meaning to your learning and being able to contextualize is important to you. You likely enjoy classes where you are able to get up, move around, and interact and collaborate with others.

Let us connect your learning style with the modalities discussed earlier. If you are more of an auditory learner, a traditional class may be a good fit. While most instructors hopefully try to engage their students and not rely on lecture, this is the class where you will get the most verbal explanations. This is also the case, possibly more, in virtual classes. In a virtual class, as it can be more difficult to engage students, instructors may lecture more. Additionally, while instructors will likely show their notes in a PowerPoint, through a document camera, or through a smart pen, it may be easier to focus on your professor's verbal explanations. However, the visual learner may appreciate the minimal distractions in a virtual class. Students can take virtual classes from the comfort of their own home, and professors often mute any background noise.

However, that does not mean the visual learner cannot benefit from a traditional class. Professors often supply a great deal of notes in class and online. More specifically, professor's frequently post ancillary notes as a PDF, Word document, or PowerPoint in an online shell or even notes with the professor narrating a video.

Is an online class a better fit for the auditory learner or visual learner? Traditionally, online classes consisted more of reading notes online, reliance on the textbook, and the student completing more work (than a traditional class). Online classes can have less distractions, as students can complete the work from a setting of their choosing. All the above favors visual learners. However, in the present day, professors generally post videos with thorough verbal explanations combined with written notes. This is why it may be worthwhile investigating the setup of a specific online class before registering.

How about the kinesthetic/tactile learner? Unfortunately, this type of learning has been marginalized in the past at the college level. However, some of the more modern modalities may help the kinesthetic/tactile learner. The emporium model, for example, may be a good fit. Students get ample opportunities to work on examples in class with minimal lecture. Additionally, the emporium class may be more relaxed, where students can move around more and take breaks as long as they complete their assignments. Some colleges are also including more coding and robotics in courses such as college algebra, trigonometry, and calculus. More specifically, students still learn the traditional math but apply various concepts to coding and robotics and utilize hands-on activities in class. The quantitative reasoning class employs lots of real-life applications and invites the opportunity for manipulatives as well, especially in topics such as probability. While teacher preparatory classes are strictly for students looking to enter the education field, these courses tend to use a great deal of hands-on activities as well.

Why in the World? Part 2

Why in the World is a Number Divided by Zero "Undefined?"

For example, $8 \div 0$ is undefined. Like multiplication by zero, this is another rule from elementary school that students were taught and told to accept. I will add that this is an important rule to master. In fact, there are many algebra, and even pre-calculus or calculus students, who will get division by zero incorrect.

Why this makes no sense: Let us hear from Andrea:

> *I never got why when we divided by zero, it was undefined. My teacher used to say, "Eight divided by zero is undefined. Mathematicians don't know what that means." And I was like "Why don't mathematicians know what that means? Are they stupid?"*

Andrea made a valid point. Mathematicians deal with complex topics such as in numerical analysis, topology, and orthogonality. How can they not understand a division concept?

Let us try to understand: Let us examine some examples through applications:

Tom brings eight cupcakes to a room with four people. How many cupcakes does each person receive?

8 Cupcakes

This translates to 8 ÷ 4, which is 2, and we can see that each person will receive two cupcakes. It is noteworthy that 8 is the dividend, which the starting amount. Four is the divisor, which indicates the number of groups in which the dividend must be split. Finally, the result, which is 2, is the quotient. Here is another example.

> Tom brings eight cupcakes to a room with only one person. How many cupcakes does each person receive?

Since there is only one person, this translates to 8 ÷ 1, which is 8. We can see the lone person gets all eight cupcakes.

Now, consider this:

> Tom brings no (or zero) cupcakes to a room with eight people. How many cupcakes does each person receive?

ZERO CUPCAKES

- Since Tom does not have any cupcakes, no one in the room receives any cupcakes. Therefore, this translates to 0 ÷ 8, which is 0. But how about this?

- Tom brings eight cupcakes into a room with no (or zero) people. How many cupcakes does each person receive?

NO PEOPLE!
Since no one is in the room, how can anyone receive a cupcake? It is impossible to arrive at an answer; therefore, this translates to $8 \div 0$. It is not even zero, because there is no one there to even receive a cupcake. That is why division by zero is undefined.

Activity

With so many options, it is important to understand your own goals and learning style before enrolling in a math class. The following questions should help.

- Do you have any thoughts about your future career goals? Is this career path in STEM or not?
- Do you see yourself enrolling in a traditional, online, or virtual math class? Why?
- How do you feel about the emporium model? Could you learn math in this environment?
- Do you prefer more of a traditional approach to learning math or an inquiry-based approach? Why?
- If you could tell your instructor in advance how you best learn math, what would you say?

References

Arem, C. A. (2003). *Conquering Math Anxiety* (2nd ed.). Brooks/Cole: Grove, CA.

Dunn, R., & Dunn, K. (1978). *Teaching students through their individual learning styles.* Pearson College Division.

Jacoby, B. (2020). *What about the other 85%.* Inside High Ed. www.insidehighered.com/views/2020/07/23/colleges-should-be-planning-more-intentionally-students-who-commute-campuses-fall

Online Learning Consortium. (2018). 2018 annual report. https://olc-wordpress-assets.s3.amazonaws.com/uploads/2019/04/OLC-2018-Annual-Report-Online.pdf

ThinkImpact. (2021). *Community college statistics.* www.thinkimpact.com/community-college-statistics/#1-community-college-vs-university-statistics

Weisbrod, B. A., Ballou, J. P., & Asch, E. E. (2008). *Mission and money: Understanding the university.* Cambridge University Press.

3

How Do I Prepare for College and Math?

Register as Early as Possible and Do Your Homework

Do not wait till the last minute to register for your math class, or any class for that matter. You will want to ensure that you get the time slot of your choosing. At this point you should know the first math class in which you need to enroll. Based on your major, you should also know your math pathway. In other words, how many math classes, after your first, do you need to take? This is something you can discuss with your advisor. Now, you need to choose a class.

In the previous chapter, I discussed the different options for math classes that schools offer. Certainly, the modality requires some research, but so does the instructor. What makes a quality math instructor can be subjective. Some faculty members may use more traditional instruction while some use more of an inquiry-based approach. Neither is good or bad, just different styles. However, you will want an instructor who is reputable for helping students learn and going the extra mile to help students learn. Additionally, you will want an instructor who is fair and consistent in his or her practices. How can you determine the best math instructor for you? Below are some options.

Choosing a Professor

A common resource for students is ratemyprofessors.com. This site consists of student reviews for that instructor from former students. The site even rates the professor on overall quality and difficulty. Additionally, students can post reviews without any kind of rigorous screening process. This site can be both reliable and unreliable. Consequently, some reviews may be done out of spite for a bad grade. Students may want to judge the validity of each review individually. Below are two different reviews regarding a professor for a college algebra class:

1)   *Too arrogant and difficult. Don't take him.*
2)   *Professor Smith's class was challenging. He moves very fast, and he gives a lot of homework. His tests can be difficult, but they are based on*

DOI: 10.1201/9781003614142-3

*everything we covered in class. There is also a lot to cover in this class, so
I guess any professor would have to move fast. Professor Smith was also
always willing to answer questions after class and during his office hours.*

Cleary the second review was more thoughtfully written and could prove
to be more reliable.

You should talk with your advisor about selecting a professor. Advisors
work with many students and have heard the good experiences along with
the bad. Advisors also work directly with faculty. Consequently, advisors can
provide objective information and can have a feel for a good match between
a student and teacher.

There may be various online sites or forums that consist of alumni from
that school. You may want to reach out to those individuals to inquire about
math faculty. Unlike ratemyprofessors.com, you may be able to gain more of
an in-depth understanding of that faculty member, and you can interact with
people and ask questions as well.

The Biggest Reason for Failure in Math and Something that May Surprise You!

What is one of the biggest, if not the biggest reason that students struggle in
a math class? It is the lack of prerequisite skills. I cannot count the number
of conversations I have had with students in which I have had to reassure
them that their struggles in math have nothing to do with their IQ or even
work ethic. They lack the foundational skills needed for that specific math
class. What makes math a challenging subject is that it is a very linear and
progressive discipline. This goes back to early elementary school. If a student has even a checkered understanding of the basic operations of addition, subtraction, multiplication, and division, he or she will struggle in
future arithmetic concepts such as operations with fractions, decimals, and
percentages. In basic algebra, if a student struggles with operations with
integers (positive and negative numbers) and rational numbers (positive
and negative fractions), he or she will struggle with evaluating expressions,
solving equations, and factoring. Additionally, students who lack essential
algebra skills will struggle in a trigonometry, pre-calculus, or calculus class.
Ultimately, in many cases, students simply have gaps, to varying degrees, in
their mathematical knowledge base. Do not just take my word for it. Here are
what some students have to say.

*I just always thought I was bad at math. For so many years, I just felt lost and
stupid in math. I never realized it was because I just had gaps. It wasn't my fault.*

Once I started filling in those gaps in algebra and arithmetic, I could learn new math. It was like a light bulb went on.

Peter

Anytime a teacher introduced something, I was lost within the first few sentences. When I went to community college, I had to take a basic algebra class. It cleared up so much for me. We started from the beginning, and I was able to finally understand things like signed numbers, adding and subtracting polynomials, and factoring. I wasn't stupid. I just needed someone to help me understand something from the beginning.

Alicia

I loved my booster class for statistics. We were able to use the time to go over things we needed for the college-level class. What was even better was we could ask questions on basic stuff. I never understood probability in middle school and high school, and I felt so dumb. I remember hearing things about "with and without replacement" and "conditional probability" and it made no sense to me. But in the booster class, we could just go over the basic stuff we needed.

Janet

I do not think this surprises many students. In fact, if you have struggled in a math class, you were likely told by a teacher or someone else that you lacked the skills needed for that class. However, here is something that may surprise you. This concept of mandatory prerequisite skills for a math class impacts everyone, not just K-12 or college students. Several years ago, the accrediting bodies, who oversee institutions of higher education, required that math faculty possess certain credentials to teach college-level math classes, and those credentials include a master's degree in math or a certain number of graduate hours in math. More specifically, for many faculty this includes a master's degree plus 15–18 graduate hours in pure math classes. There were several college math faculty members who did not have such credentials. Generally, those faculty members had taught lower-level math classes. However, those teachers who did not have the required credentials were required to return to graduate school to complete their requirements, and some of those teachers struggled.

As part of a study, I spoke with several college math teachers who needed to complete graduate-level math classes. Some shared horror stories. Professor Holton noted that he enrolled in an online program to complete his graduate-level math classes. His highest math class had been differential equations, an undergraduate math class just above the calculus series. However, Professor Holton's first graduate class was abstract algebra. Basically, he went from a sophomore or junior-level math class to a graduate-level math class. What made matters even worse was that it had been over 15 years since Professor Holton even took a math class. Consequently, Professor Holton did not make it through the class, as he was completely lost and frustrated. He even received a "25%" on his first and only exam!

Think about this. Professor Holton, like others, is an accomplished and intelligent person who taught math for a living! So, why did Professor Holton and other math instructors struggle in their graduate-level math classes? Because they were lacking prerequisite skills. To be successful in graduate-level math, a person must at least have a thorough understanding of all undergraduate-level math. For many of these math instructors, it had been years, or decades, since they completed a math class, and, like Professor Holton, they had significant gaps in their mathematical knowledge base. Consequently, these instructors felt lost in these graduate classes and doubted their own ability. Does that sound familiar?

Being Behind May Not be Your Fault

If you have ever entered a math class underprepared, you may have felt this was your fault. In fact, you may have believed that lacking the prerequisite skills was just another example of being a bad student. This is not always true. A student may be exposed to bad instruction or face external difficulty during a math class, and consequently, he or she may fall behind in class. This can lead to mathematical gaps, and the student is then unprepared for the next math class.

The COVID-19 pandemic wreaked havoc on the entire planet but also on the educational system. The timelines varied between states, but students were out of school and had to learn virtually or completely online for one to two years. Consequently, due to the lack of in-person instruction, many students fell behind in their studies.

> *I was never a good math student, but that year and 3 months during COVID really screwed me up. I had to learn algebra on Google Classroom, and I had such a hard time learning that way. I felt like I had to teach myself a lot. I was failing miserably, but I passed. I think a lot of other kids were failing pretty bad as well, but we passed, because what was the teacher going to do? Fail us all? Problem was I went from being bad at math to being completely lost.*
>
> Jim

Jim was not alone. Research has shown that students throughout K-12 across America are an average of one year behind in their math progress. This means that students have developed gaps in their mathematical knowledge base, and these gaps only worsen and widen as they progress in grade levels. It is especially disheartening as this impacts students all the way down to early elementary school. Waiting to address prerequisite skills will only make matters worse. Find out the skills you need and start addressing them now! We will address how to do this in the next section.

Preparing for Your Math Class Even Before the First Day

So, how can you prepare for your first math class, or any math class, before it has even begun?

> *I was so scared about going back to school as an adult, so before my first math class, which was algebra, I looked into what material I needed. My school's tutorial services were so helpful. They gave me material that was covered early in the class, and I started watching videos, doing worksheets, and working with tutors. That helped my anxiety so much. I felt so much more confident going to class the first day. It also helped, because it put me ahead of the class.*
>
> Dina

> *The best thing I ever did was start to prepare in advance for my math class. I tested into college algebra with the booster class, and I failed it twice. So, before the third time, I did some research on what topics I needed for the class. I found some videos on those topics, like factoring and finding the slope of a line, and some worksheets online, and I started practicing. It helped me so much.*
>
> Bob

The best way to attack a math class is head-on. Prior to your first math class, conduct some research on which topics you will need to be prepared for the class. The best way to do this is to contact your professor or another professor and ask about these topics. You can also contact your college's tutorial services. Once you have determined these topics, here are some suggestions:

- Locate some online videos on these topics, but you need to be careful. Anyone can post videos on YouTube or similar sites. You will want quality videos with accurate information. Khan Academy includes several videos hosted by Sal Khan. These videos include content from arithmetic through calculus. Brian McLogan is a high school math teacher who posts several quality videos in algebra, trigonometry, and pre-calculus. Mario DiBartolomeo is an experienced math educator who posts excellent videos in algebra. It is noteworthy, however, that while many of Mario's videos are free, some come at a cost.

- As noted earlier, talk with the people at the tutorial services center at your school. Most colleges have a centralized tutoring area. Introduce yourself to the people who work there and inform them that you are looking to get some advanced help with a specific class. They may have some resources to help you. These are people you will want to get to know early in your college career.

- Some colleges even offer head start programs where students can spend time before classes start (e.g., end of the summer) working on the math they will need and focusing on prerequisite gaps. Ask the people in the tutorial center if your college offers such opportunities. Ask your advisor as well.

- Ultimately, spend some time practicing the math you will need. Let me ask you a question. In the remaining days before your math class starts, is your time better spent worrying about your class and if you will succeed or spending the time working on the math you will need and building up your confidence? In Chapter 1, we discussed the issue of control and how feeling out of control causes anxiety for many people. Getting a head start will allow you to feel in control and you will begin to confront your fears of math gradually and at your own pace.

To further help you prepare for your math class, Figures 3.1–3.5 provide general prerequisites with sample questions that you need for five general math classes. Keep in mind that these prerequisites may vary between schools. However, depending on your introductory class, this could give you a baseline of topics to review. The answers to these questions are in Appendix A.

Course: Introduction to Algebra or Pre-Algebra

Typical Topics you Need to Understand Entering this Class:

- Addition, subtraction, multiplication, division with whole numbers.

  1) Add: $456 + 898$ 2) Subtract: $7,009 - 4,999$

  3) Multiply: 453×31 4) Divide: $3,412 \div 23$

- Round the following numbers to the specific place value:

  1) 234.566 (tenth)   2) 768.098 (hundredth)

- Addition, subtraction, multiplication, division with fractions.

  1) Add: $\dfrac{3}{5} + \dfrac{1}{6}$ 2) Multiply: $\dfrac{3}{8} \times \dfrac{8}{21}$ 3) Divide: $\dfrac{6}{7} \div \dfrac{18}{35}$

FIGURE 3.1

Prerequisites for Introduction to Algebra/Pre-Algebra.

- Addition, subtraction, multiplication, division with decimals.
  1) Add: $5.6 + 2.456$ 2) Subtract: $92.3 - 79.998$
  3) Multiply: $(3.45)(43.2)$ 4) Divide: $4.34 \div 0.05$

- Solving proportions.
  1) Solve: $\dfrac{5}{x} = \dfrac{10}{3}$
  2) If a jogger runs 2 miles and burns 185 calories, how many calories would he burn jogging 3 miles?

- Percentages
  1) Find 34% of 75. 2) What percent of 60 is 32?
  3) Ann bought a dress that was marked down by 35%. If Ann paid $110.00 for the dress, what was the original price?
  4) Convert to a percent and round to the nearest tenth of a percent. 0.04578

- Addition, subtraction, multiplication, and division with signed numbers.
  1) Add: $-12 + (-36)$ 2) Subtract: $-36 - (-51)$
  3) Multiply: $(-12)(-10)(-8)$ 4) Divide: $-108 \div 9$

Please note: While a scientific calculator can perform all the above operations, you can make it easier for yourself by being able to compute these types of problems without a calculator.

FIGURE 3.1 (Continued)

Statistics Can be Deceiving

When reviewing Figure 3.5, there are not many prerequisites for an introductory statistics class, especially compared to college algebra. Students primarily need a solid background in arithmetic and some algebra, and keep in mind that many colleges offer co-requisite or booster classes in statistics. So, introductory statistics must be easy, right? Not so fast!

Introductory statistics is a very fast-paced and deep class. While the content is obviously different, I have colleagues who have compared the pacing of a statistics class to a calculus class. More specifically, students learn new

Course: Elementary or Intermediate Algebra

Typical Topics you Need to Understand Entering this Class:

- **An understanding of arithmetic concepts.**
  See example in Figure 3.1.

- **Addition, subtraction, multiplication, and division of signed numbers.**
  See example in Figure 3.1.

- **Evaluating algebraic expressions.**
  Evaluate $4x^2 - 5y^3$ if $x = -3$ and $y = -5$.

- **Using the distributive property.**
  Distribute the following: $-3(4x - 12)$.

- **Simplifying algebraic expressions.**
  1)   Add: $(4x^3 - 3x^2 - 6x + 2) + (6x^2 - 10x - 9)$.

  2)   Subtract: $(12x^2 - 10x + 20) - (25x^2 - 20x - 53)$.

- **Solving basic linear equations.**
  1)   Solve: $3x - 5 = 9x - 15$

  2)   Solve: $\dfrac{3}{4}x - 5 = \dfrac{2}{3}x + 6$

- **Solving algebraic word problems.**
  1)   The second of three numbers is 1 less than the first. The third number is 5 less than twice the second. If the third number exceeds the first number by 12, find the three numbers if their sum is 68.

  2)   One weekend Bill earned three times as much as Jim. Tom earned $5 more than Jim. In all, they earned $60. How much did each earn?

FIGURE 3.2

Prerequisites for Elementary or Intermediate Algebra.

Course: College Algebra or Pre-calculus

Typical Topics you Need to Understand Entering this Class:

- An understanding of the topics listed in Figures 3.1 and 3.2.
- Solving complex equations with fractions and parenthesis.

  Solve: $\dfrac{2}{3}(x-2) = \dfrac{3}{4}(x+5) - 3$

- Factoring expressions.

  Factor the following:

  1) $x^2 - 10x + 9$ 2) $x^2 - 16$ 3) $4x^2 - 8x - 32$

  4) $4x^2 - 7x - 11$ 5) $x^3 + 27$

- Solving quadratic equations by all three methods.

  1)   Solve by factoring:

       $3x^2 - 13x + 10 = 0.$

  2)   Solve by completing the square

       $x^2 - 8x - 48 = 0.$

  3)   Solve by using the quadratic formula:

       $2x^2 - 5x - 9 = 0.$

- Finding the slope and equation of a line.

  1)   Using the points $(-4, 5)$ and $(-5, -7)$ find the equation of the line.

- Solving systems of equations:

  Solve: $\begin{cases} 2x + 3y = 5 \\ 4x - y = 7 \end{cases}$

- Simplifying rational expressions:

  Subtract: $\dfrac{x+1}{x-4} - \dfrac{x-1}{x^2 - 7x + 12}$

- Multiplying and dividing rational expressions:

  1) Multiply: $\dfrac{x-7}{x^2 + 3x} \cdot \dfrac{x^2 + 10x + 21}{x^2 - 49}$ 2) Divide: $\dfrac{2x^2 - x - 1}{2x^2 - 3x + 1} \div \dfrac{1}{4x^2 - 1}$

- Solving inequalities:

  Solve:

  1) $3 - 2(x - 3) \le -4(x - 5)$ 2) $\dfrac{1}{3}(x - 5) < -3(x + 8)$

For pre-calculus, it is possible you may need to know some trigonometry. I included some examples in Figure 3.4.

FIGURE 3.3

Prerequisites for College Algebra or Pre-Calculus.

Course: Calculus

Typical Topics you Need to Understand Entering this Class:

- A deep understanding of all the topics in Figures 3.1, 3.2, and 3.3.
- Understanding functions: polynomial and exponential functions.
- 1) What is a function?

  2) Consider the following function: $f(x) = x^3 - 2x^2 - 3x + 5$
  As $x \to \infty$ $f(x) \to$ __.
  3) Solve: $4^{2x-1} = 16^{3x+2}$

- **Solving logarithmic equations using the properties of logarithms.**
  Solve: $log_2 x + log_2 (x-3) = 2$

- **A solid understanding of major trigonometric concepts:**

  1)   What is the reference angle for $\dfrac{7\pi}{6}$?

  2)   Consider the following function: $f(x) = 3\sin(2x - 4\pi) + 1$

       Find: a) the amplitude, b) the period, c) the horizontal shift, and d) the vertical shift

  3)   Rewrite the following as one trigonometric function with no fractions:
  $$\frac{1 + cot(x)}{1 + tan(x)}$$

  4)   Rewrite $\sin\left(x - \dfrac{5\pi}{6}\right)$ in terms of sine and cosine.

- **Finding asymptotes:**
  Find the equation of the vertical and horizontal asymptotes for the following functions:

  1) $f(x) = \dfrac{3x}{x^2 - 8x + 12}$ 2) $f(x) = \dfrac{4x^2 - 1}{x^2 + 3x - 28}$

FIGURE 3.4

Prerequisites for Calculus.

Course: Introduction to Statistics

Typical Topics you Need Entering this Class:

- **An understanding of all arithmetic concepts (see Figure 3.1).**
- **An understanding of mean, median, mode, and range.**
  1) Find the mean, median, and range for the following numbers:
      34, 56, 89, 90, 23, 45, 53, 36, 80, 20, 10, 35, 79, 90, 92.
  2) Find the mode for the following numbers:
      40, 50, 40, 43, 45, 50, 40, 50.

- **Finding the slope and equation of the line.**
  See Figure 3.3.
- **Solving linear equations.**
  See Figure 3.2.
- **Evaluating algebraic expressions.**
  See Figure 3.2.

FIGURE 3.5
Prerequisites for Statistics.

concepts often and then must apply such concepts to additional new concepts. However, do not take my word for it.

> I liked statistics; it was the first math class that really made sense to me, but the class goes so fast. There is no let up. If you slack off, you will fall behind, and it's so hard to catch up.
>
> Jerry

> Everything starts off basic, but it gets hard really fast. Like probability, we started off with basic probability, and that was good, because I really didn't understand much about probability, but then it got really complicated really fast.
>
> George

To further illustrate my point, consider two probability problems from introduction to statistics.

Problem 1:
A fair die is rolled one time. What is the probability of rolling a 5?

Problem 2:
A company manufacturing electronic components for home entertainment systems buys electrical connectors from three suppliers. The company prefers

to use Supplier A because only 1% of those connectors prove to be defective, but Supplier A can only deliver 70% of the connectors needed. The company must also purchase connectors from two other suppliers, 20% from Supplier B and the rest from Supplier C. The rates of defective connectors from B and C are 2% and 4%, respectively. You buy one of these components, and when you try to use it, you find that the connector is defective. What is the probability that your component came from Supplier A?

Even if you have never studied probability, it should be obvious that "Problem 2" is much more difficult, and it takes a lot of time, work, and mastery to get from "Problem 1" to "Problem 2."

What if I Struggle with Basic Arithmetic?

Do you struggle with concepts such as fractions, decimals, and percentages? How about adding, subtracting, multiplying, and dividing whole numbers? You are not alone. Incoming college students have struggled in arithmetic dating back to the early 1700s. This was so much the case that Yale University implemented a proficiency exam that focused on arithmetic (Arendale, 2002).

Struggling in arithmetic concepts can be both daunting and demoralizing:

> *I was always in the stupid math as long as I could remember, and it made me feel stupid. I could never get fractions. Heck, I struggled with multiplication and division.*
>
> Cindy

> *I was afraid to go back to school because of math, but it was more than just math. I couldn't do really basic math. How was I going to pass college math if I couldn't get basic math?*
>
> Emily

Can students who struggle in arithmetic eventually pass a college-level math class? Yes, but it is important to choose the correct pathway and fill in the needed gaps. At one point, colleges offered quarter or semester-long courses that focused solely on arithmetic. However, these courses proved to be unsuccessful, and due to financial aid issues and pressures from individual states, most colleges have eliminated these courses. However, there are options if you struggle with arithmetic. Some colleges offer truncated arithmetic review classes. These may be week-long classes that offer a brush up in arithmetic concepts. This is something worth looking into.

If you are concerned about your arithmetic skills, you should also speak with the people in the tutorial services at your college. Is there anything they can do to help you with arithmetic skills? Is there an individualized program where you could work with math tutors? Summer is a great time for this. Enrollment at most colleges is lower during the summer, and the people at a college's tutorial services may have more time and resources during the summer.

If you feel deficient in basic arithmetic, there is an additional option. There are state-funded adult basic education programs that are free to students. Such programs allow students to focus on basic reading and math skills. Some colleges even work closely with these programs. Your advisor may have more information regarding such programs. However, students who enroll in these adult basic education programs do not receive college credit. These programs have proven to be advantageous for students before attempting a pre-algebra class at a college. However, some states have limited their offering of adult basic education programs.

How is Your Number Sense?

"Number sense" is a term thrown around mathematics. There are varying definitions, but number sense refers to a student's ability to understand, relate, and connect numbers. Number sense, or lack thereof, can serve as either a major advantage or disadvantage for students in math. A student with a strong number sense possesses basic mathematical ability without being calculator dependent. Additionally, a student with a strong number sense can answer a mathematical question and have a basic idea as to whether it is correct. Ultimately, a strong number sense can spearhead a student through courses such as algebra, trigonometry, pre-calculus, and beyond. Conversely, a lack of number sense is a major reason students struggle in math. Number sense can develop and evolve. As students get stronger in math, their number sense becomes stronger. If you are testing into an introductory math class such as a developmental math class, quantitative reasoning, introduction to statistics, or even college algebra with a booster class, please see Figure 3.6 for a brief quiz to help you understand your own number sense. If you are in a higher-level math class, your number sense is likely strong. Again, these are very basic math concepts, but see how well you can do without a calculator. The answers are in Appendix A.

Give yourself one point for every correct answer, and if you scored below a "9," you may want to practice similar problems to help develop your number sense.

1. How many times does 12 go into 232 (yes, there is a remainder)?
2. What are all the numbers that divide evenly into 72?
3. What is the biggest number that can divide evenly into both 12 and 20?
4. What is the lowest number that both 8 and 10 can divide evenly into?
5. Come up with the pairs of numbers that multiply to 48.
6. Come up with the pairs of numbers that multiply to 64.
7. Nine times what is 63?
8. What is one half of 68?
9. What is three times as much as 32?
10. What is one-fourth of 60?
11. What number times itself gives you 121.
12. Using addition, subtraction, multiplication, and division, construct a problem that gives you 42 as an answer.

FIGURE 3.6
Number Sense Quiz.

Developing a Sense of Community and Belonging

Whether or not you loved or hated high school, you were a part of a community. You were with the same people for several hours a day for at least four years, and by the time you graduated, it likely felt like a family. You may not like your job, but working with the same people day in and day out provides a sense of community. So, how can you develop a sense of community in college?

Starting college, especially if you do not know anyone on campus, can be intimidating and can feel isolating. Students who reside at a college or university can develop a sense of community and belonging in the dorms. If you make a connection with your roommate or people in your dorm, you will start going to dinner and other events together. Generally, there are freshmen social events where you can meet people.

Community college, however, can be isolating. What I remember about my first day of community college was that I did not speak to another person, except to ask for directions. I went to my three classes, and then I went home. This is unfortunately the case for many students. They go to class and go home, barely speak to anyone, and never develop any sense of community. Of course, living at college does not guarantee that students will develop a sense of community. You may not bond with your roommate, or you may feel like you do not fit in at school.

Developing a sense of community and making connections with other people have been shown to help students academically. Therefore, it is important to research your college or community college to determine how you could fit in and develop a sense of belonging. Colleges typically offer a variety of clubs and even social events.

> *I love to play chess. I've always loved chess, but it was always something I did by myself on my phone or laptop. I couldn't believe that my college had a chess club. I loved our meetings, and I made so many new friends. It made me want to go to school.*
>
> Derek

> *I knew I wanted to be a nurse, but I was really scared about my classes and if I could do it or not. My community college actually has a club for future nurses. I was able to talk to people who were already in nursing classes. More importantly, I was able to talk with people who had passed the dreaded math classes. It made me feel more confident that I could do it.*
>
> Joanne

> *Going back to school as an adult can feel isolating, but I joined this club; it was a student government club for adults returning to school. It was a way for us to have a say in how we could improve the school, and we also organized charity events as well. It was a great way for me to meet new people and not feel so alone. I remember us sharing our anxieties about being older and going back to school and taking math! I'm so grateful for this club.*
>
> Dina

My Own Story in Developing a Sense of Community

As I mentioned earlier, I entered community college with a great deal of math anxiety, as I had been a subpar student in high school. However, I also entered community college with social anxiety disorder. I wanted to connect with people, but I did not know how, and my social anxiety was a major barrier. With the advice of my parents, I decided to try and join a club. I wanted to succeed academically, but I also wanted to feel a sense of belonging. I attributed, in part, my poor academic performance in middle school and high school to my feeling like a social outcast. Therefore, I knew if I could make connections with other people, I had a better chance academically, as I would be more content.

Joining a club with social anxiety disorder is easier said than done. I needed a club that was small and informal. I always had an interest in writing, so when my English professor mentioned that she hosted a writer's group every week, I decided to attend. I really enjoyed the intimacy of this group, and I began to attend each week. Not only did I develop my writing, but I met new people. I made such an impression that the group's advisor asked me to be the student senate representative, which opened even more doors for me. I continued to join more clubs and attend more events and meet more people. By my second year, I could not walk a few hundred feet in any direction on campus without greeting several people or stopping for a conversation. Joining my first club set me in the right direction both academically and socially.

The Challenges of Meeting People

Why is meeting or getting to know new people more difficult in the present day? Walk into any room of people, anywhere, and what will you see? Are they socializing with each other? Probably not. Their heads are probably down, and they are glued to their phones. Additionally, many of these people are wearing ear buds, further blocking out the rest of the world. Sadly, this is what you will likely see when you walk into your first class. This is why it is imperative to try and join a club or attend a social event that is of interest to you and in which you feel comfortable. Chances are the people there will want to socialize with others! Additionally, when you enter class the first day or first week, try and sit near people who seem friendly and open to chatting with one another.

Ensuring You Have Your Materials

Will you need your math textbook or access to some type of math software on the first day? The answer to this question varies. Some professors will not expect students to have their course materials on the first day, while others will want you to come to class with your materials on the first day. Most colleges no longer have brick-and-mortar bookstores. They utilize online resources. You can check through your college's online bookstore for the materials you need. However, you can also send your professor a polite email a week or so prior to the start of the semester. Below is an example.

> *Hi Professor (insert name). My name is (insert name), and I am excited to take your math class this semester. I would like to be as prepared as possible. Could you let me know if I need my textbook or any course material on the first day? Thank you, and I look forward to working with you.*

It would be a wise idea to order your materials as soon as possible. There are times where students who use financial aid experience a delay in receiving textbooks.

Disability Services (If Needed)

Several students have learning differences, and in many cases, these differences can serve as a barrier to learning math. If you have received accommodations in the past for a learning difference or suspect you may have a learning difference, it is imperative to contact your school's office of disability services. Learning differences include dyslexia, dyscalculia, auditory processing disorder, and language processing disorder, and students can receive help.

> *I have something called dyscalculia. I think it used to be called math dyslexia. Basically, it affects the way I process and retain numbers. I always thought I was just an idiot. But I learned it had nothing to do with my intelligence. Basically, I record my professor's class. I take notes and work with a counselor to help understand my notes. I get extra time on exams, and most importantly, I get a quiet place to take my exams.*
>
> Jerry

Unfortunately, there many students, like Jerry, who struggle in math but are unaware that they have some type of learning difference. Do not wait till you are already struggling in your math class. Find out if you have an issue in advance and start receiving help!

Are Your Basic Needs Being Met?

The term "starving students" has been around for decades. The term refers to the fact that college students are often poor. Yes, I remember those days. I recall driving around in a beat-up Toyota Camry with about 150,000 miles on it. While the car made all kinds of strange noises, I remember thinking, "Please don't breakdown; please don't breakdown." Additionally, I remember making $6.85 per hour (late-1990s' money) and having very little money in my checking account. While in community college, I decided to rent a room in a boarding house to be near the college. It was a single 100 square foot room without a sink, and I shared a bathroom with two other people. That is not exactly the lap of luxury, but it is a much better lifestyle than many current college students endure.

Food insecurity is defined as the lack of consistent access to food. Unfortunately, around 40% of United States college students face food insecurity to some degree (Selcho, 2022). Food is a physiological need. As humans, if we do not have enough of it, there is very little we can accomplish, and that certainly includes obtaining a college degree. If you find yourself contending with food insecurity, there are people and places that can help. For example, most colleges contain food pantries. These food pantries distribute non-perishable foods for free to people in need.

Food insecurity and housing insecurity often go hand in hand. Just as many college students face food insecurity, many students are homeless. In fact, there are 1.4 million undergraduate students who are homeless (McKibben, Wu, & Abelson, 2023). You cannot attempt a math class, let alone a college degree, if you are facing food or housing insecurity. You need to address this immediately. There is funding available for students who face these issues. Here is a story from one student who faced homelessness and food insecurity.

> *I was at the point where I was living in my car, and I would have to go through trash cans to get food. I would take showers at the college in the locker room and handwash some of my clothes in the locker room sink with soap. I never asked for help, because I felt ashamed. I finally did [ask for help], and I was shocked. A counselor helped me get funding for a temporary place to live and a job. I don't know what I was thinking. There was no way I could succeed in any class if I didn't have a place to live.*
>
> Jerry

Most college students need to complete an introductory psychology class. In that psychology class, you will learn about Maslow's Hierarchy of Needs. The theory states that we need to have our physiological needs met, and such needs include having enough food and water before we can accomplish additional goals. We also need to feel safe and have secure housing. See Figure 3.7 for a full image of Maslow's Hierarchy of Needs.

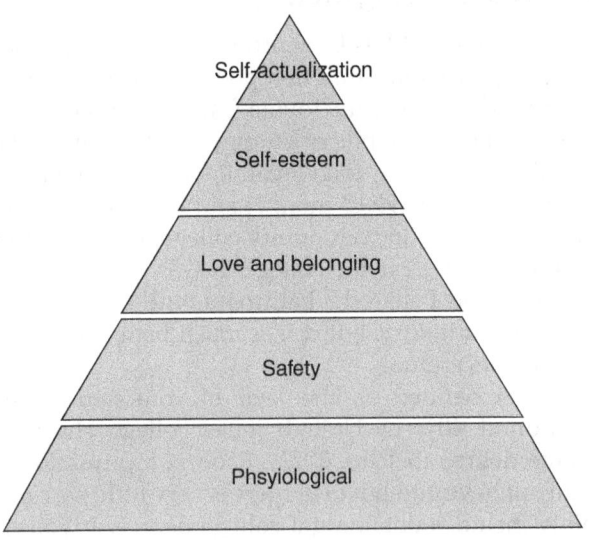

FIGURE 3.7
Maslow's Hierarchy of Needs

If you are facing food or housing insecurity, note this on your FAFSA form, if you are applying for financial aid. Additionally, you need to speak with someone on your college campus who can help with this. Before attempting a math class, you must ensure that your basic physiological and safety needs are being met.

Figure Out Your Math Time!

Poor time management is another ingredient for lack of success in a math class. I will discuss attendance later in this book, but before you even attend a single class, you need to ensure that you have time outside of class to devote to your math studies. The process of mastering a math topic does not simply take place in the classroom. You need to complete your homework assignments and practice problems outside of class. Each week, you should devote at least six to eight hours outside of class for your math homework and for practicing math.

Let us be honest. Our time feels a resource that other people continue to try and steal. Outside of school and work, it can feel like family and friends are constantly seeking our time. Therefore, you need to be proactive. What hours each day can you devote to "math time"? Each morning from 7–8 am? How about some days from 6–7 pm and other days from 10–11 pm? "Math time" needs to protected time. This is time that you will devote to your math classes, and no one can interrupt this time. Yes, you need to schedule this time in advance before the semester has begun and before you even see your math class schedule. Trust me, you will make use of this time, whether it is completing homework or simply reviewing your notes or working on practice problems. Think about your math time, because I am going to ask you to define it in the activities section.

A Fresh Start

Before taking a math class or even starting college, keep something in mind. There are events that have happened in your life that may make you feel anxious, insecure, or inferior, but remember that no one at your college is judging you based on your past. Your professors do not read your previous transcripts or speak with people who knew you in the past.

As mentioned earlier, I was not a good student in high school, and I always felt that was my reputation. I brought home disappointing report card after disappointing report card. The outcome from all my parent–teacher conferences was always the same, bad. I expected to fail simply because that was who I was.

Starting college was a transformation for me. I decided to buckle down and work hard. For my math class, I placed into basic concepts of algebra, and while I was terrified, I was determined to succeed, because I wanted a better life for myself. I decided to stay on top of my studies in math, and I devoted 1–2 hours per day in the school's tutorial center. I completed all my homework assignments in a timely manner, and I asked for help when needed. I completed my basic algebra class with an "A." An interesting turn of events began happening during my freshman year. Other students began asking for my help with their studies and began treating me as the smart guy. By the end of the school year, I was tutoring several students. Additionally, I was having a conversation with my basic algebra professor, and I mentioned that I was apprehensive about taking college algebra. She said, "I wouldn't worry. You are very smart, and you work hard." I could not believe she was talking to me. So, wipe your slate clean. This is your new beginning.

Why in the World? Part 3

Why in the World Doesn't "PEMDAS" Always Work?

Many of us remember this in some form or another. We learned it around fifth or sixth grade. It is the order of operations. For example, when given a problem with multiple operations such as:

$$9 + (6 - 2)^2 \cdot 3 \div 8$$

we were told to use the order of operations to simplify the problem and arrive at an answer. The order of operations instructs us to do the parentheses first, the exponents second, multiplication third, division fourth, addition fifth, and then subtraction last. One way to remember the order of operations is PEMDAS, which is the first letter of each word in the correct order. Some may also remember Please Excuse My Dear Aunt Sally as a way to retain this information. Following the order of operations, and simplifying each part one step at time, in the above problem, we get:

$9 + (6 - 2)^2 \cdot 3 \div 8$, start by simplifying the Parentheses.

$9 + 4^2 \cdot 3 \div 8$, then simplify the Exponents (recall the 4^2 means $4 \cdot 4$)

$9 + 16 \cdot 3 \div 8$, then Multiply.

$9 + 48 \div 8$, then Divide.

$9 + 6$, then Add.

Our answer is 15, and there is no need to Subtract.

Why this makes no sense: However, now consider the following problem: $15 - 5 + 2 - 3$. If we were to follow PEMDAS, we would add 5 and 2 to get 7. Then, we would subtract $15 - 7 - 3$ and get 5. Is that correct? Keep in mind the above problem tells a story.

There were 15 people in a room; five people left; two more walked in, and then three left.

If you follow the story from beginning to middle to end, or left to right, we see that there are nine people left in the room, not five.

Let us try to understand: When addition and subtract appear together in the problem, we must work left to right. Therefore, the "AS" in PEMDAS can be misleading. The same holds true for multiplication and division. We must work left to right.

The following problem appeared on Facebook, inviting people to arrive at an answer:

$6 \div 3 \cdot 2(4 + 6)$. There were several different answers; however, a common incorrect answer was 10. This was because people followed PEMDAS, in order, and multiplied the 3 and 2 before dividing. However, since both multiplication and division appear in the same problem, we need to multiply and divide from left to right. Therefore, our answer is 40. Of course, the funny part on Facebook is not so much the varying answers but people yelling at each other, defending their answers! The bottom line is that PEMDAS is a way to remember the order of operations, but it does not take into account the basic rules of adding and subtracting and multiplying and dividing from left to right.

Activity

- What math class do you need to take? What are the topics you need for this class? Have you practiced some problems on these topics?
- What clubs or organizations at your college or community college interest you? Why?
- Do you have adequate food or housing? If not, have you taken the steps to secure that?
- What hours on which days of the week will be your "math time"?

References

Arendale, D. (2002). *History of supplemental instruction (SI): Mainstreaming of develop-mental Education.* Center for Research on Developmental Education and Urban Literacy, General College, University of Minnesota.

McKibben, B., Wu, J., & Abelson, S. (2023). *New federal data confirm that college students face significant- and unacceptable-basic needs insecurity.* The Hope Center: Temple University. https://hope.temple.edu/npsas

Selcho, M. (2022). *An estimated 40% of U.S. college students experience food insecurity.* Deseret News. www.deseret.com/2022/10/26/23424063/how-many-college-students-face-food-insecurity/

4

Getting Off and Staying On the Right Foot

A major part of combatting anxiety is preparation and being proactive to an upcoming situation. The first three chapters of this book have hopefully helped you prepare for college and your math class. In this chapter we will examine how to get off to a good start and maintain the course.

What is Your Math Personality?

People have math personalities! Answer yes or no to the questions from Figures 4.1 and 4.2 and count which personality you answered "yes" to more.

Several years ago, I had two very good students in my intermediate algebra class. Both participated actively in class, had a great work ethic, and performed well on exams. However, they drove each other crazy. Stacey became irritated with Jim because when I introduced a new topic, he wanted to discuss real-life applications, engage in discussions, and even go off on tangents. Jim would also want to investigate different ways of completing one problem. I could predict Stacey rolling her eyes every time Jim asked "how?" or "why?," which was quite often. Stacey wanted a step-by-step process for each topic. When Jim would go off on a tangent, she would raise her hand with frustration and say, "OK, so what was step 3?" or "What came after step 5?"; Jim, on other hand, would get annoyed that Stacey interrupted his inquisitive thinking with process-based questions.

The reality is that, just as in life, students have different math personalities. Math personalities tend to be quantitative or qualitative. Quantitative thinkers tend to be more structured, prefer a linear problem-solving process, and focus more on the solution to a problem rather than the process. Qualitative thinkers prefer flexibility, relying more on intuition, and focus more on the process than a solution.

DOI: 10.1201/9781003614142-4

1) Do you like it better when a math problem has an identifiable pro-cedure, recipe, or method?

2) Do you have difficulty when the problem you are working on doesn't have a set formula?

3) Do you perform better at computational parts of exams as opposed to word problems?

4) Do you perform adequately on quizzes but have difficulty with unit exams? Or do you perform adequately on unit exams and have difficulty on a final?

5) Do you have difficulty when you automatically have to rely on math concepts from the past in order to do a current problem? For example, using fractions or decimals to solve an algebra problem?

6) Do you find it difficult when going to another math class and another professor shows a different method for solving a math problem than the one you have become accustomed to? In other words, you like to stick with one way only of solving a specific problem.

7) Do you view math a multi-storied building; you must finish one level to move on to the next?

8) Do you prefer the teacher to explain how to complete a math problem as opposed to working together in a small group with your classmates to complete it?

FIGURE 4.1
The Quantitative Personality.

Is one personality or way of thinking a better fit for math? The actuality is that traits from both personalities are imperative, and the key is to achieve balance. The study of math relies on algorithms. That is, there is a proced-ural process to solving mathematical applications. Of course, procedural processes are not confined to math. Most fields in the sciences use algorithmic or procedural processes to solve applications. You may be familiar with the scientific method of an experimental design that starts with a hypothesis or theory and ends with a conclusion of the study. However, success in math also involves critical thinking. Students must be able to rely on intuition and focus on the process without fixating on the solution. Relating math to real-world concepts, known as contextualization, is imperative in developing a deeper understanding of math.

Why is understanding your math personality important? This may explain some of your previous struggles in math. Your math personality may be another reason why you did not mesh with a previous teacher's pedagogical

1)   Rather than formulas or set procedures, do you look for connections, similarities, or patterns?

2)   As an example, in geometry, do you need pictorial representations (need to draw the pictures) as opposed to just memorizing the formulas? In fact, do you struggle with memorizing formulas or a step-by-step process?

3)   Do you question when you will ever apply the math you are learning to real life, as opposed to just taking what you are taught and accepting it?

4)   Do you enjoy class discussions on "the best way to solve a problem"?

5)   Do you view math like a puzzle; do you look for connections and try to connect all the related ideas from the past and the present?

6)   Do you like it when your professor shows more than one way to solve a math problem?

7)   Have you ever been told by your teacher or others that you "over think" math?

8)   Do you prefer to work in small groups to try and solve a problem rather than being guided through by your teacher?

FIGURE 4.2
The Qualitative Personality.

style. There is nothing wrong with being more quantitative or qualitative. However, now, that you know your math personality, use this as an opportunity to be more open-minded in learning math. In fact, another myth in math is that some people cannot learn math through algorithms. More specifically, qualitative thinkers are at a major disadvantage. If you are more qualitative, keep up the critical thinking, but keep an on open mind on learning algorithms. Remember, you use step-by-step processes every day of your life and across the disciplines. For example, dance and math may seem like night and day. However, isn't dance a series of steps to create a desired movement? That is an algorithm. Dancers must master algorithms! If you are more quantitative, yes algorithms are necessary in math, but also keep in mind that focusing too much on algorithms may become problematic when you must solve a similar but different problem. Be open to new ideas and even new ways of solving a problem. Try to visualize how you may use this mathematical application in everyday life. The balance of both traits will help you become a better math student. Later in this chapter, I will present an application from basic algebra and how the two different personalities may approach this topic.

The First Day

Since starting kindergarten, over 40 years ago, I have been through many "first days." There were all my first days to start each school year in my K-12 and college years. Then, there have been my first days of teaching elementary, high school, and college. The first day of school is still anxiety-ridden for me. Yes, after over 25 years of teaching, I still feel nervous on the first day of school before I greet each class. I want to do my very best to reach all my students; however, with so many unknown personalities, I have no idea if I am reaching them the right way. I am also questioning if I can get these students to where I need them to be, academically. My point, it is perfectly normal to be nervous on the first day of school, especially starting college.

My primary concern on the first day is making a good impression on my students. I want them to know that I care about them and will go the extra to help them, but that this will be a rigorous class requiring hard work. Students have several anxieties on the first day, but one of them is making a good impression and getting off on the right foot. As the saying goes, "you never get a second chance to make a first impression." Remember, the first day is just one day. Good or bad, the first day does not define your college career. Nonetheless, there are steps you can take to try and ensure a good start.

> *Get to class early. In fact, go to campus a couple of days before your classes start and practice walking around campus going to your classes. Know where your rooms are.*
>
> David

> *I hate the dead silence on the first day of class. Everyone is just sitting there awkwardly waiting for class to start. People have their heads down and are looking at their phones. Some people have earbuds in. Try and make eye contact with anyone around you and just smile.*
>
> Dina

> *The first day is a lot of going over the syllabus and the course expectations. But you might go over some math. If your professor asks you a question, be courageous and raise your hand and answer. You'll make a good impression, and it will break the ice.*
>
> Jessica

> *On your way out of class, make eye contact with your professor and just say "thank you." Or stop by and say "thanks, and I am looking forward to your class."*
>
> George

What if you are taking your math class online? What is a first day like, and how can you get off to a good start?

> *An online math class can be overwhelming. Try to focus on the introductory stuff. Read the syllabus but also see if your professor posted an introductory video. Look at all the expectations like homework and exams and day-to-day stuff. Take a deep breath and make note of anything you don't understand. Make a list and right away nicely email your professor or even visit during an online office hour.*
>
> Abby

> *Taking an online class can be isolating. It feels like you're going at it by yourself, but think about it, you're part of an entire class going through the same struggles and trying to accomplish the same goal. There is usually some kind of general place to post in the online forum. Introduce yourself to the class and invite other students to do the same. It will break the ice a little. If you don't know where to post this, ask your instructor.*
>
> Melinda

As a math professor, I agree with all the students' suggestions. I will add that you will need to remind yourself not to get overwhelmed. Hearing about the expectations from your math class combined with the expectations from your other classes may increase your anxiety. Again, remember that it is only Day 1 of your journey. Also, keep in mind that you have already set up your "math time," that is hours each week devoted to your math. In the activities section, I am going to provide questions to ask yourself after the first day of class.

Technology in Math

If you are coming to college directly out of high school or it has been a short time since you left high school, chances are your high school math teachers employed various kinds of technology in your classes. If it has been a while since you were in high school, the technology in your college math classes may be overwhelming at first. In this section, we will examine how you may see technology employed in your class and how you can use technology to help you in math.

Calculators

Handheld calculators (four function, scientific, and graphing) have been around since the latter part of the 20th century. Therefore, the use of calculators in the college classroom should not be a shock to anyone. However, different colleges and even departments within colleges use them differently. From the 1990s and well into the 21st century, there were intense debates regarding where calculators, of any kind, should be used in developmental math classes. One group asserted that developmental math students should develop number sense, and master algorithms and formulas without the calculator to develop a strong mathematical foundation to best prepare for college-level math. The other group posited that it is unfair to prohibit the use of the calculator since students have used it for so many years (high school and before). Additionally, this group has argued that people use calculators in everyday life, so they should be used in the college classroom.

While some colleges still prohibit the use of the calculator in lower-level math classes, most colleges have moved toward allowing some type of calculator usage in all math classes. In lower-level math classes, such as an

- Adding, subtracting, multiplying, and dividing fractions.

  1) Add: $\dfrac{7}{8}+\dfrac{1}{3}$ 2) Subtract: $\dfrac{7}{10}-\dfrac{2}{3}$ 3) Multiply: $\dfrac{6}{11}\cdot\dfrac{22}{23}$

  4) Divide: $\dfrac{8}{17}\div\dfrac{51}{56}$

- Computing multiple operations with fractions:

  $$\frac{3}{5}+\frac{2}{3}-\left(\frac{1}{4}\cdot\frac{4}{5}\right)$$

- Converting a decimal to a fraction and vice versa.

  1) Convert to a decimal and round to the hundredth place: $\dfrac{7}{11}$.

  2) Convert to a fraction: 0.875.

- Converting a mixed number to an improper fraction and vice versa.

  1) Convert the following improper fraction to a mixed number: $\dfrac{54}{7}$

  2) Convert the following mixed number to an improper fraction: $6\dfrac{3}{8}$

FIGURE 4.3

Can you do these on a calculator?

arithmetic brush-up or a pre-algebra class, students may only be permitted to use a four-function calculator. A four-function calculator only computes operations of addition, subtraction, multiplication, and division, so students may need to understand the rules of fractions and additional complex arithmetic formulas. Most iPhones contain four-function calculators. Be prepared however, as some math classes require a graphing calculator. Depending on the type of graphing calculator required, it may cost at least 110–120 dollars.

If you have used at least the scientific calculator in high school, using it or the graphing calculator in college should not be too much of a challenge. However, if it has been a while or you have never used either a scientific or graphing calculator, either one can be intimidating. In fact, learning math and the nuances of a calculator can increase math anxiety. Again, being proactive is a great way to combat anxiety. Prior to starting your class, you may want to practice some basic operations (see Figure 4.3 and see the solutions in Appendix B) with the scientific or graphing calculator. Of course, as I discussed in Chapter 3, you will also want to be able to perform these problems without a calculator as well. Knowledge of these basic operations can help acclimate to the scientific or graphing calculator and put you at ease somewhat as you start your class.

Interactive Math Software

In Chapter 2, I discussed the use of interactive math software. Again, examples of interactive math software are MyLab Math (aka MyMathLab), Hawkes, or ALEKS. These programs are utilized in online classes, but even traditional classes utilize these programs as homework assignments. While using these programs, students answer math questions online and receive an answer immediately. Generally, if students answer incorrectly, they can make more attempts or try similar problems. Oftentimes, the problems are accompanied by a short video or guided practice. For each unit, students generally must complete a set of assignments (e.g., 5 sets of 10 or 15 problems) before each exam.

Some high schools utilize interactive math software, but some do not. Additionally, students who have been out of school for some time may be unfamiliar with this type of program.

> *I was out of school for 25 years, and when I was introduced to MyMathLab on the first day of class, my heart was beating so fast, and I felt short of breath. I was like, "I'm scared enough about math, but now I have to do math on the computer!" My professor started going over what we needed to do in MyMathLab, and I just wanted to get up, walk out, and never go to school again.*
>
> Donna

However, Donna has some advice for students who may feel the same way.

> *Just go home the first day and register for the program. It's like registering for any kind of online subscription. You'll just need to give them your information, and basically you come out of it with a username and password that you'll need to remember. Just take it a step at a time and get registered. Then, try a few math problems out. You'll feel better. It sounds kind of scary on the first day, but you'll get used to it.*
>
> Donna

George added some recommendations:

> *If you run into any technical issues, don't panic. I used Hawkes for my first class, and I had trouble registering. I emailed my professor right away, and we met on Zoom. It was something I was entering in wrong, and we got it fixed. Just take care of it right away. The kids in my class who ran into problems waited way too long to start working on Hawkes.*
>
> George

I will add that your professor will likely provide you a schedule of when the online assignments are due. The key is not to fall behind. Now, you know what you will be working on during most of your allotted math time!

Zoom Meetings

While college professors have traditionally offered office hours (set times where students can meet with the professor), many professors now offer office hours through Zoom. This has long been the case for online classes, but now many professors are willing to meet with students virtually even in traditional classes. This can prove to be convenient for students who cannot make it to campus for an office hour.

I'm in My Nightmare! I'm Lost in Class!

Part of the fear of taking a math class is that all too familiar feeling of sitting in class while the professor lectures and simply feeling lost. As hard as you try, you just can't understand the material, and you feel like you are barreling toward failure. Again, you are not alone.

> *I was so afraid to take math in college. I just hated being in a math class and feeling so confused and stupid. I would start crying when thinking about it.*
>
> Tatiana

That's exactly what I was afraid of. Sitting in class not knowing what the [expletive] was going on and being so scared my teacher would call on me and everyone would think I'm stupid.

James

You know what's really annoying. When you're following along with the teacher somewhat, and all of a sudden, he says something, and you're completely lost. I hate that, but it happened to me all of the time.

Nick

Can this happen in an online math class? Absolutely.

The first time I tried taking an online class, I remember the first couple of topics going all right, but once we got into the second week, I was just lost. My professor would post a video online and online notes, but I just couldn't follow all the way through. I just started panicking because I knew I had to understand the material, but I just wasn't.

Ana

Let us take a dive into why this scene is so anxiety-ridden.

- As I mentioned in Chapter 1, a loss of control can lead to an increase in anxiety. If students cannot follow the instructor and comprehend the material, they are not in control of the situation.
- When events in life go too fast, we can feel anxious. Oftentimes, I find myself asking out loud if I can just hit a pause button!
- Past trauma makes this worse. Students are already familiar with this helpless feeling. It is like a reliving a nightmare.
- Especially in an online class, students can feel very isolated when struggling. In a traditional class, other students may be asking questions or expressing their frustrations. Online students feel as though they are going at it alone.
- Students know the potential end results of this situation, which could be failing the class. This could escalate concerns such as students falling behind in their plan to graduate, losing financial aid, scholarship-funding, or being academically dismissed. Those thoughts could cause panic for anyone.

Let me provide a harsh reality. This scene will very likely happen. However, it is important to remember that all of us, even those who have advanced degrees in mathematics, have felt lost in a math class. The question therefore is not "is this going to happen to me?" It is "How will I deal with feeling lost in a math class the first time this happens?" Here are some tips for both traditional and online learning. Refer to Figure 4.4. Do not panic if you do not know how to solve this problem. It is just an example for a multi-step problem.

Solve: $\frac{2}{3}(x-4) = \frac{1}{4}x + 5$

Step 1: Remove the parenthesis by applying the distributive property:

$\frac{2}{3}x - \frac{8}{3} = \frac{1}{4}x + 5$

Step 2: Remove the fractions by multiplying each term by the lowest common denominator:

$(12)\frac{2}{3}x - (12)\frac{8}{3} = (12)\frac{1}{4}x + (12)5$

$8x - 32 = 3x + 60$

Step 3: Move the variable to one side of the equation:

$8x - 32 = 3x + 60$

$-3x \ldots\ldots\ldots -3x$

$5x - 32 = 60$

Step 4: Isolate the term with the variable:

$5x - 32 = 60$

$+32 = +32$

$5x = 92$

Step 5: Isolate the variable:

$\frac{5x}{5} = \frac{92}{5}$

$x = \frac{92}{5}$

FIGURE 4.4
Multi-step problem for solving equations with fractions.

- Take a deep breath and remind yourself that this is just one class. You may feel lost right now and you may not catch on for the rest of the class, but you still have control over how you handle this going forward.
- Take careful notes but outline what you do not understand. If it is a multistep problem, put a star next to what is giving you trouble. Try to stay away from the generic "I do not understand" and narrow your thoughts to "I do not understand this step" or "I do not understand how the professor is going from this step to that step." For example, regarding Figure 4.4, is it Step 1 or Step 2?
- Visit your college's tutorial center and review your issue with a tutor. Most colleges now offer virtual tutoring where you can meet with a tutor via Zoom or some other forum. Additionally, see your professor during office hours to get help.
- Ask your tutor or your professor if your issues have more to do with prerequisite skills or new material. If your issue has more to do with a prerequisite skill, you may need to practice some problems on that topic. For example, if you are struggling with Step 1 (in Figure 4.4), you may need to review using the distributive property with fractions. If you are struggling with Step 2, you may need to review concepts such as finding the least common denominator or multiplying fractions by whole numbers.
- For both traditional and online classes, consider working with other classmates.
- Keep in mind that students who allow these issues to escalate run into trouble. Their math gaps will only widen, and they will only become more lost.

Does Reading a Math Textbook Help?

Whether it is in PDF form or a traditional paper copy, most math classes utilize a math text. I still recall taking a workshop, my freshman year, to help with college success, and the presenter began one session in a monotone voice saying, "How to get the most from your math textbook." I zoned out after that. I never had success from a math textbook, and I had no faith in that workshop changing my mind. In this section, we will unpack whether a math textbook helps or hinders success.

Most teachers will instruct students to read their math textbook. This is sound advice, as the material in the textbook relates directly to the content covered in class. However, students often complain that the information presented in the textbook is difficult to follow. Why is this?

- Textbooks will often show the steps to a sample problem; however, they may not address prerequisite skills. More specifically, the author(s) will transition from one step to another without discussing the prior knowledge needed for a problem.
- Author(s) often use technical and academic language, as opposed to explaining problems in laymen's terms. This may be because math textbook authors are mathematicians who have difficulty breaking the content down to simpler terms.
- Oftentimes, the explanations in the textbook's sample problems may not apply to all the problems in the activity section or online problems. More specifically, there may be problems in the activity section that are not covered in the sample problems.
- Textbooks favor the visual learner and may marginalize the auditory and kinesthetic/tactile learner.

So, is a textbook a good resource for students or not? It can be. In fact, for more advanced classes such as calculus, students should be able to read and understand textbook explanations. In higher-level math classes students need to be able to comprehend advanced and technical language. However, for students who are in lower-level classes and struggle with math anxiety, there are other resources available, which are discussed in the next section.

What Can You do Outside of Class to Help?

Let us state the obvious. Good attendance, as in your job and as in life, is imperative in a math class. It is also important to keep up with homework assignments. Missing too many classes and not keeping up with the material is an ingredient for failure. Additionally, in Chapter 7, we will break down how to study for a math test. But what else can students do outside of a math class to achieve success?

Become a Regular at the Tutorial Center

Tutorial centers exist for your benefit. In fact, if students did not use these facilities, there would be no tutorial centers. Some tutorial centers offer open hours where students can simply come in and work on their math and receive help when needed. If that is the case, make it a point to go in there and simply work on your homework. This way you can simply receive help when a problem arises.

Some college's tutorial services work by appointment only. You may need to sign up in advance to work with a tutor. This may be for an hour or two each week. Again, you may be able to do this either in person or virtually. Sign up for this time in advance. Plan the questions you will ask your tutor before the session.

Study Groups

Working together with your classmates is a great tool for success. For a traditional class, it may be awkward to approach other students in person and inquire about a study group. However, you could ask your professor to make an announcement for students to sign up for a study group. In an online class, this is even easier. You can post a note in the general forum and ask if anyone wants to form a study group or even multiple study groups. For everyone's convenience, you can meet via Zoom and discuss your struggles. In a traditional class, you can meet either in person or online. Students have reported that working together through a math class relieves math anxiety and can increase confidence. You will see that others are struggling with the same material that you are. There will also be content in which you are stronger but others struggle, and you can help other students. Additionally, explaining and teaching the material to someone else is a great way to get better at it! In fact, a great professor once told me, "the best way to learn something is to teach it."

Online Resources and Artificial Intelligence

As mentioned in Chapter 3, there are online videos available in most math topics. These videos especially help the auditory learner with clear and concise verbal explanations. Such videos will also assist the visual learning with some clear and concise written steps. Moreover, these explanations are short and to the point and are in layman's terms without a lot of technical verbiage. Kinesthetic/tactile learners will especially benefit from interactive software such as MyLab Math, as students can practice several problems. Even when getting a problem correct, students can continue to practice similar problems.

The use of artificial intelligence (AI) seems destined to impact education, and there are ways students can utilize AI to help them. ChatGPT is a mostly free resource that can serve as a supplement to help student learning. An account for ChatGPT is easy to set up. Once you are in, you can type in a message such as "I need help with subtracting polynomials." I listed my question along with ChatGPT's response in Figure 4.5. The explanations likely benefit the visual learner; however, the explanations are in simpler

I need help with subtracting polynomials.

ChatGPT said:

Absolutely! Subtracting polynomials involves distributing a negative sign and combining like terms. Here's a step-by-step guide:

Example problem

Subtract the following polynomials:

$(3x^2+4x-5) - (2x^2-3x +7)$

Step 1: Distribute the negative sign

Rewrite the problem by distributing the negative sign across the second polynomial:

$3x^2+4x-5-2x^2+3x-7$

Step 2: Combine like terms

Now, combine terms with the same degree (i.e., x^2 terms, x terms, and constant terms):

1. For x^2 terms: $3x^2-2x^2 = x^2$
2. For x terms: $4x+3x=7x$
3. For the constants: $-5-7=-12$

Step 3: Write the simplified expression

So, the answer is:

$x^2+7x-12$

Would you like to try another problem? Or let me know if you have questions about any of these steps!

FIGURE 4.5
ChatGPT Example 1.

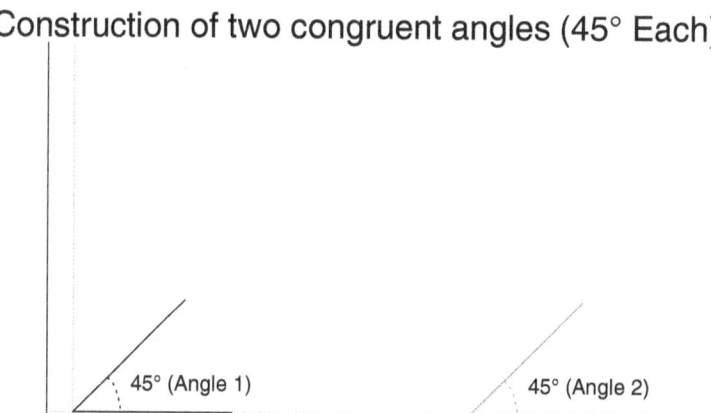

FIGURE 4.6
ChatGPT Example 2

terms than the traditional textbook, which can benefit everyone. Simpler terms, after all, can alleviate anxiety. AI resources such as Alexa and Siri will provide an answer to a math problem; however, the aforementioned resources will verbally provide the step-by-step process to a problem, which may not benefit all students. AI resources such as Alexa or Siri can be helpful when looking up a math definition of some sort. For example, "Siri, what is a coefficient in laymen's terms?"

Keep in mind that AI can make mistakes. In fact, ChatGPT even lists the disclaimer "ChatGPT can make mistakes. Check important info." Additionally, not all the ChatGPT's explanations may help everyone, and it may depend on the wording. I asked ChatGPT, "Construct two congruent angles." We will discuss angles in Chapter 5, but congruent angles are two angles of equal measure. Figure 4.6 shows ChatGPT's response, and this visual seems clear and understandable.

However, I asked ChatGPT to "Show two congruent angles" and Figure 4.7 shows the response. This visual could confuse students. It is interesting how "Construct" and "Show" could lead to two vastly different responses.

What Happens When Life Happens?

Earlier, I stressed the importance of good attendance and keeping up with homework assignments as imperative ingredients for success in a math class. Failure to do so often results in a lack of success, but let us face it, life happens, and life can be unforgiving. First off, all of us get sick. As of

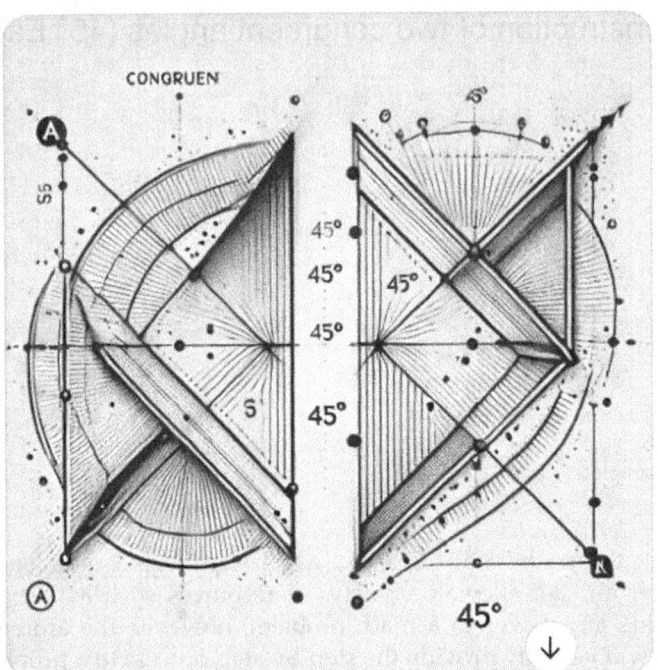

FIGURE 4.7
ChatGPT Example 3

this writing, I have had COVID three times, just as an example. When I was lying in bed, completing work responsibilities was not my priority; I needed to recover. Additionally, family responsibilities can take students away from school responsibilities as well. For example, students may need to care for a loved one.

Math classes can move at a fast and unforgiving pace. So how do students keep up when "life happens"?

- First off, this is why it is imperative to keep up with your assignments and your studies and not fall behind. If you are already behind, further absences will make matters worse.

- When you are aware that you will be absent, contact your professor and explain the situation. Find out what sections you are covering on that day. If you are ill, you may simply need time to recover. If you can, try to learn the material. Some professors post notes or even videos for each section. Find online videos that cover this content.

- If you are still struggling with the new content, upon returning, visit your college's tutorial center to work on some problems and visit your professor during office hours. Remember, tutoring or visiting your professor through Zoom may also be an option.

- An online class offers more flexibility, but there are still deadlines. Therefore, you can still fall behind just as easily as in a traditional class.

- I will discuss this more in Chapter 8, but there are times where you may need to withdraw from a class due to outside circumstances. This may be the case if you need to take a prolonged absence from class, or if it is an online class, you may not be able to keep up with the assignments. Unfortunately, this happens in real life. For example, people need to take extended leaves from their jobs due to various situations.

I Understood What We Were Doing in Class, But I Drew a Blank When I Tried It at Home

It used to drive me crazy. I could follow along what we were doing in class. And if I was working with my teacher or my tutor, I could get it, but when I tried it on my own, I couldn't do it. It's like I could do any math problem as long as I had my tutor or my teacher chained to me.

Julie

This is true for many students, and it can be quite frustrating. However, it is important to remember that this is good. After all, understanding math as it is explained is much better than feeling completely lost. This means you can get it; you simply need practice. A mistake many students make is they assume that because they understand it in class, they do not need to practice outside of class, or they wait too long to practice.

Think about this. One hour after class (or after learning new material), our brains can only recall 50–60% of what we learned. After 24 hours, our brains only recall 30% of what we learned, and that percentage continues to decrease. That is why practicing math every day is so imperative to ongoing learning. This can help the student retain 80–90% of the material. More specifically, do not wait several days to cram in an online assignment; spread the assignment(s) over several days with practice each day during your "math time."

Frequent practice can help, but let us address the issue of drawing a blank if your teacher or tutor is not present. This is where taking notes in class is imperative.

I know this sounds really stupid, but I used to not take notes in my class, or even when I took notes, I wouldn't look at them when I did my homework. No wonder I was lost!

Julie

Taking accurate notes in a math class is imperative. You can then use these notes as a crutch when you are completing your homework assignments. All too often, I have witnessed students enter college unaware of how to take notes. If you feel you struggle with notetaking, feel free to show your notes to your professor and ask for any tips on better notetaking. For an online class, do not assume that notetaking is unnecessary. Yes, your professor will likely have notes for you in some form, but as you view the lectures or notes, take notes in a way that makes sense to you.

For traditional classes, many students benefit by recording the classes. This is especially beneficial for the auditory learner. As you work through problems on your own, hearing your professor's voice should help. Again, if there is part of a problem in which you are getting stuck, get help immediately.

Math Vocabulary: Yes, that Sounds Awful!

While the term, "mathematics" strikes fear in many people, the term, "vocabulary" strikes a sense of boredom in most people. Most students remember learning vocabulary in elementary school, which basically refers to learning new words. Of course, this was generally followed by the dreaded vocabulary tests where students had to define a word and then write it in a sentence. Understanding math vocabulary basically refers to being able to define and understand terms that refer to mathematical concepts (e.g., coefficient, variable, exponent). The combination of fear and boredom does not usually end well. However, the student's comprehension of math terminology is vital for his or her success, so we need to address this.

Every one of us has been in this situation. You are involved in a conversation, and you are following along. However, the other person starts using words that you do not understand, and you are thinking, "If I could just understand what this word meant, I could follow along and even contribute to this conversation!" Does that sound familiar? Same situation in a math class. The teacher may explain content, but various words keep you from understanding the entire lesson. It could, and likely has, gotten worse for you. You are trying to complete a homework assignment or even an exam, but certain words keep you from moving forward.

That used to drive me crazy. Things like conditional probability or independent or dependent events [in probability] used to just stop me. Once I saw those words, it was like I hit a wall.

George

I could never understand the difference between an expression and an equation, and I felt too stupid to ask.

Audrey

Like missing prerequisite concepts, not understanding math terminology creates gaps in the knowledge base, which in turn leads to struggles. As a great English professor once told me, "You do not know what a word means until you can define it."

Asking questions in a face-to-face class is a necessity. As soon as your professor mentions a term which you are unfamiliar with, ask! Raise your hand and when called on say something like, "I'm sorry if you mentioned this before, but what does a (whatever the word is) mean?" I guarantee at least one of your classmates will be grateful to you as they had the same question. If you feel uncomfortable asking out loud, see your professor right after class, or send an email with terms that you do not understand. You can see your professor during an office hour. In an online class, if you come across terms that you do not understand, email your professor immediately. Figure 4.8 is an example of an appropriate way to email your professor with a question on math terminology or any subject matter.

See Table 4.1 for an explanation on some basic mathematical terms in arithmetic and algebra to get you started. We will explore more advanced terminology in Chapter 6.

Dear Professor XXX

I hope you are well. I just have a question regarding the current lesson. I'm just having some trouble understanding what a function is, and I want to make sure I have a good idea of what is it is before I move forward. Could you explain it to me?

Thank you,

Student Name

FIGURE 4.8
Inquiry to the professor.

TABLE 4.1

Introduction to Basic Terms

Term	Definition	Example
Numerator	The top of a fraction	$\frac{3}{5}$: 3 is the numerator
Denominator	The bottom of a fraction	$\frac{3}{5}$: 5 is the denominator
Equivalent fractions	Two fractions that are equal but have different values	$\frac{1}{2} = \frac{3}{6}$
Factor	A factor divides into a number with no remainder. Therefore, the factors of a number are those numbers that divide into that number	The factors of 24 are: 1, 2, 3, 4, 6, 8, 12, 24
Reduced fraction	A reduced fraction is in simplest form. There is no number that can divide into both the numerator and the denominator	$\frac{2}{3}$ is a reduced fraction. $\frac{6}{10}$ is not a reduced fraction. We can reduce it by dividing the numerator and denominator by 2: $\frac{6}{10} \div \frac{2}{2} = \frac{3}{5}$
Multiple	Multiples are obtained by multiplying a number by a certain number various times	The multiples of 6 are 6, 12, 18, 24, 30, 36… This is because $6(1)=6, \ 6(2)=12, \ 6(3)=18$ and so forth
Variable	A letter that represents some quantity	This could be most letters in the alphabet, but "x" is the most common variable. In $4x - 6$, x is the variable and represents some number or quantity
Coefficient	The number in front of a variable	In $6x^2 + 5x + 2$, both 6 and 5 are coefficients
Constant	The fixed number that does not change and does not have a variable attached to it	In $6x^2 + 5x + 2$, the 2 is a constant
Expression	A combination of numbers, variables, and operations	$7x - 5$
Equation	A statement that two quantities are equal. There will be an equal sign	$7x - 5 = 16$
Term	A single number, variable, or combination of numbers and variables separated by an addition or subtraction sign	In the expression, $8x^2 - 5x + 2$, the terms are $8x^2, -5x$, and 2

TABLE 4.1 (Continued)

Term	Definition	Example
Monomial	An expression with one term	$12x^2$
Binomial	An expression with specifically two terms	$x-2$
Polynomial	Any expression with two terms or more	$11x^2-6x-10$
Simplify	Rewriting an expression so that there are no like terms	We can simplify the expression: $9x+8y+10x+4y$ By adding and subtracting the coefficients of the like terms and get: $19x+12y$

Suggestions for Each Learning Style

Let us examine some suggestions as to how each learning style can make the most in a math class.

Tips for the Auditory Learner

Remember, you learn by talking and listening. It may help you to record your professor's guided practice in class and play it back while completing your homework. When completing a problem, talk yourself through it. In fact, pretend that you are explaining it to a struggling student. You can even use a song to help you memorize various facts. For example, in graduate school, when studying for a major exam, I had to memorize what various academic experts said regarding the history of higher education. I used the song, "According to You" by Orianthi to help. The lyrics to the song begin with "According to you/I'm stupid. I'm useless; I can't do anything right." So, I altered the lyrics to, and just picture me singing, "According to Jacobs (2012), since 2007, more than a dozen states cut funding for developmental education." Hey it worked!

Tips for the Visual Learner

You learn best by reading and writing. Design flashcards and use them for terms that you need to memorize. You may need to color code your notes to highlight different parts of a problem. It will help to write out problems from your notes over again while studying. For each class, write out a few sentences attempting to summarize what you have learned that day.

Tips for the Kinesthetic/Tactile Learner

How can the kinesthetic/tactile learner stay on the right path? As mentioned earlier, the kinesthetic/tactile learner benefits from interactive learning, practice, and movement. In an online class, kinesthetic/tactile learners should move around as much as possible when completing math problems or lessons. Pace around the house and play music while you complete your math assignments. Also, once you are introduced to a problem, try another one right away. A traditional class can be a challenge for the kinesthetic/tactile learner. Kinesthetic/tactile learners have bodies that are more restless. Additionally, the professor may introduce multiple sample problems before you can work one on your own. Remind yourself that you will soon be able to practice problems on your own. Also, take advantage of any breaks that your professor provides. If your professor utilizes group work, take this opportunity to stretch and move around the room. Even if your professor transitions from guided practice to the students completing problems on their own, stand up and stretch. When you are at home, practice your math through as much movement as possible. Jump on a trampoline, for example, when trying to understand different math terms. Like the auditory learner, you can make up songs or even sing them when it comes to various formulas or algorithms. For example, the Internet has several songs on the quadratic formula.

Kinesthetic/tactile learners also benefit from understanding how math is related to real-life applications. Do not be afraid to try and understand how material that you are learning relates to real life. For example, consider the equation $400x = 10,000$. This is a basic algebra problem where we are supposed to solve for "x." Here is what this means, however. We are trying to save $10,000 to buy a car. So, we are going to save $400 each month, and $400 multiplied by the amount of months will lead to the amount of money saved, which again should lead us to 10,000. We are trying to determine how many months it will take us to save $10,000 if we save $400 each month, so we divide each side by 400 to isolate x: $\dfrac{400}{400}x = \dfrac{10,000}{400}$. We get 25; therefore, it will take us 25 months to save $10,000 if we save $400 each month.

Back to the Quantitative and Qualitative Thinkers

Since we have discussed some mathematical terms, I want to circle back to the quantitative and qualitative thinkers with a specific example. This topic is covered in basic algebra, so students could see this in middle school or in a developmental college math class. Here is a basic example:

Simplify the following: $4x + 3x$.

The quantitative thinker will want to understand the algorithm, and the algorithm states to add or subtract the coefficients and keep the terms. Therefore, the answer is $7x$. However, the qualitative thinker will want to understand why this is and see patterns and visuals. Consequently, this is basically saying that we have 4 of something and we are then adding 3 of that same quantity. The variable, "x" stands for whatever the quantity is. Let us say that "x" represents apples. How about 4 apples plus 3 apples?

That is 7 apples.

Why can't we add $4x + 3y$? Because that is like adding 4 apples with 3 bananas. Can you add?

No, the answer is simply 4 apples and 3 bananas or $4x + 3y$. Here is another problem:

Simplify: $10x + 12x^2 + 6x + 15x^2$

The quantitative thinker will focus more on the algorithm, which again states to add or subtract the coefficients of the like terms and keep the terms. However, this can be tricky, because x and x^2 are so similar. The answer is: $16x + 27x^2$. However, students may try to add all the coefficients and arrive at an incorrect answer such as $43x^2$. Consequently, the quantitative thinker may want to tap into some qualitative thinking. Recognizing that x and x^2 are two different but similar terms, perhaps allow x = books, while x^2 = magazines. So, there are 16 books and 27 magazines when combining like terms.

Why in the World? Part 4

Why in the world is a negative times a negative a positive?
For example, let us multiply $(-120)(-3)$, which falls under the rules of signed numbers. The answer to this is 360.

Why this makes no sense: How could multiplying two negatives result in a positive? Wouldn't a negative multiplied by a negative just give us a really big negative?

Let us try to understand: We need to return to some basic rules of signed numbers, and signed numbers include both positive and negative numbers. We use negative numbers when dealing with owing money or very cold temperatures. How about adding or subtracting signed numbers, such as $-11+(-15)$? If Mike already owes 11 dollars and adds on a debt of 15 dollars, Mike now owes 26 dollars or –26. How about if Mike owes 18 dollars but then receives 10 dollars? Mike still owes 8 dollars. Isn't that $-18+10 = -8$? That is why adding and subtracting signed numbers have two rules.

1) When the signs are the same (both positive or both negative), we add the numbers and keep the sign: $-11+(-15) = -26$.

2) When the signs are different, we subtract the numbers and take the sign of the higher number: $-18+10 = -8$.

The rules of addition and subtraction at least make some sense when related to money. However, let us return to the ambiguous rules of multiplication.

You may remember, or not, that the rule for multiplying signed numbers is that when the signs of each number are different [e.g., $(-120)(3)$], the sign of the answer is negative. Therefore, the answer is –360. How about we start with why $(-120)(3) = -360$? Let us say that you join a health club for $120 for 3 months. More specifically, you have paid $120 for each month. That is –120 + (–120) + (–120), which adds up to $360 that you have paid, and the amount subtracted from your bank account is –$360. Of course, it is much quicker to rely on the rule, which again is, when multiplying different signs, the answer is negative. But how on Earth is a negative multiplied by a negative a positive? Consider this; you move away but forget to cancel your health club membership, and of course the health club takes $120 from your checking account each month for the first three months. However, you realize the mistake and contact the health club and prove that you have moved. Consequently, the health club decides to reimburse you for the three months of fees. Let us consider what has happened here. The health club negated the three months of payments, and you received $360. That is a double negative leading to $(-120)(-3)$, which again is 360.

Practice Some Math

Earlier in this chapter, we introduced math terminology. If you have been away from math for some time and need a review of basic algebra and some arithmetic, here are some practice problems. A student's mastery of these concepts is also imperative going forward to higher math.

1) Is $7x^2 - 5x - 2$ an expression or an equation? Why?
2) For $13x^2 - 9x - 5$
    a) Identify the coefficients and the constants.
    b) Identify the terms.
3) Simplify: $10x^2 + 15x + 32x^2 + 20x$.
4) Solve: $5x - 12 = 28$.
5) Give your own example of a monomial and a polynomial.
6) List the factors of 84.
7) List the factors of 96.
8) List the first 10 multiples of 8.
9) List the first 10 multiples of 6.

Activity and Math Exercises

- What is your math personality? How has your math personality impacted your previous math classes? Based on your math personality, what are some of your needs in your college math class?
- This is difficult, but imagine yourself in class (in-person or online), and you hit a barrier where you simply do not understand something. Panic and anxiety start to set in. How do you handle this?
- What is your experience with technology in math? Are there certain math videos or software which you feel comfortable with? Or does technology in math intimidate you?
- In this chapter, we applied the three types of learning styles and how you can acclimate to a math class. Again, what is your learning style and what are some helpful techniques to help you get off on the right foot to mastering the math content?
- List various strategies you can use to be successful in math class.
- List various actions that will likely lead to failing a math class.

5

I Hate, Hate, Hate These Math Topics!

We have addressed that many students fear and have a disdain for mathematics. However, what are the topics that make students hate math? Why do students despise these topics so much? We will answer those questions in this chapter, plus provide some tidbits on how students can conquer these topics. Word problems are a despised topic, so much so that they get their own chapter. Consequently, we will address word problems in the next chapter. For now, let us focus on some other detested math topics. At the end of this chapter, I will provide some practice problems for each of these unpopular topics. Feel free to do as few or as many as you wish whenever you are ready.

Fractions

This should be no surprise. Many students of all levels despise fractions. In fact, there is generally a collective groan in any math class when a teacher utilizes fractions of any kind. In most schools the concept of fractions are introduced in early elementary school. However, it is in around fourth or fifth grade that students begin learning about the operations (addition, subtraction, multiplication, division, reducing) of fractions.

Why Do Students Hate Fractions?

I just always hated fractions. I just always got confused when I was supposed to get a common denominator or reduce or change to multiplication. It was just a bunch of rules that I never understood.

Alicia

Fractions are just symbolic of why I hated math, just rules that I never understood about a topic [fractions] that I never understood.

Bob

DOI: 10.1201/9781003614142-5

> *I really hated dividing fractions. Why are you supposed to flip the second fraction*
> *and multiply? Because some mathematician said so?*
>
> Derek

Fractions are an important part of life because they are used in everyday life. After all, fractions represent a non-whole quantity, and life does not consist of merely whole quantities. As evidenced by the testimonials above, many students do not understand the idea behind fractions when they are introduced, and moreover, the algorithms behind addition, subtraction, multiplication, and division add immensely to their confusion. In fact, Jean Piaget, renowned psychologist, has posited the stages of childhood development. Specifically, Piaget has labeled ages 7–11 as the concrete operational stage and age 12 and older as the formal operational stage. What does this have to do with learning fractions? Without seeing physical examples of how fractions work, the algorithms likely make no sense to students. Additionally, the leap from the simplicity of operations with whole numbers to the complexities of fractions can be overwhelming and intimidating for many children. As we discussed in Chapter 1, early experiences in math that are bad can shape a student's outlook of math. Making matters worse, fractions do not just go away. They are utilized in algebra and beyond.

> *One thing that was so frustrating in algebra was I would understand something,*
> *but as soon as the teacher put in a fraction, I was lost.*
>
> Patrick

Where Will I Need Fractions?

Review the prerequisite figures in Chapter 3. Fractions are everywhere! No matter which math pathway in which you endeavor, whether it is algebra, trigonometry, calculus, quantitative reasoning, or statistics, you will see fractions. Again, this is because fractions are so much a part of real life. Students who need to complete college algebra will contend with fractions more extensively than those who need to complete quantitative reasoning or statistics; however, all students attempting a college-level math class should be reading to tackle fractions.

What Can We Do About Fractions?

This is a tough one. When reading opinions from "experts" in math education, many will say that fractions, along with other mathematical topics, are taught incorrectly in elementary school, and teachers should utilize more hands-on manipulatives so that students truly understand the meaning of fractions. Other "experts" argue that it is imperative students master the algorithmic process. The truth is that both sets of "experts" are correct. Students should understand the meaning of fractions, but they also need to master the algo-rithmic process for higher-level math. However, this does not help the stu-dent preparing for college math who fears fractions, so let us focus on them. Adult learners have had exposure to fractions both in school and in real life. However, I would recommend watching some elementary videos that intro-duce the basics of fractions. For example, Kahn Academy's "Introduction to Fractions" would be a good start. In the next section, we will look at some tips to understanding the basics of fractions.

What Should I Know about Fractions? And Some Tips to Get Started

Students attempting any college-level math class need to have a thorough understanding of fractions. Additionally, students should be able to under-stand the algorithms of fractions but also compute complex fractions on the calculator. More specifically can you perform $\frac{2}{3}+\frac{1}{4}-\left(\frac{2}{3}\cdot\frac{1}{4}\right)$ on the calcu-lator? The answer is $\frac{3}{4}$.

A Look at Adding and Subtracting Fractions

First Example: $\frac{1}{3}+\frac{1}{3}$. If a pie were to be cut into three slices, and you were to eat one slice in the morning ($\frac{1}{3}$ of the pie) and one slice in the afternoon (another $\frac{1}{3}$ of the pie), you would have eaten $\frac{2}{3}$ of the pie. The numerator represents the part of the pie we have eaten. The denominator represents the number of equal parts to make up a whole. What did we just do? We added the numerators and kept the denominator. That is the rule for adding or subtracting fractions.

Second Example: $\frac{2}{3}-\frac{1}{4}$. What is the difference from the first example? The denominators are different. How can we keep the denominator if the denominators are different? We need to create two new but equivalent fractions that have the same denominator. How can we do that from a 3 and a 4?

- First, we need to find the lowest number that both 3 and 4 can divide into. This is known as the least common multiple. Let us count by multiples of 3 and 4:

  3, 6, 9, 12, 15...

  4, 8, 12, 16, 20...

Notice how 12 is the lowest number that both 3 and 4 can divide into. Therefore, 12 is our least common multiple, our least common denominator.

- Next, we need to create two equivalent fractions with a denominator of 12. How can we do that?

$$\frac{\dfrac{2}{3} - \dfrac{1}{4}}{\dfrac{}{12} - \dfrac{}{12}}$$

- Now, we need two new numerators. How do we do that? We divide each old denominator (3 and 4) into 12. When dividing 3 into 12, we get 4. We multiply 4 by the old numerator of 2 and get 8! For the second fraction, we divide 4 into 12 and get 3. We multiply 3 by the old numerator of 1 and get 3! So:

$$\frac{\dfrac{2}{3} - \dfrac{1}{4}}{\dfrac{8}{12} - \dfrac{3}{12}}$$

Now we can subtract: $\dfrac{8}{12} - \dfrac{3}{12}$, which is $\dfrac{5}{12}$.

How Does This Apply to Higher-Level Math?

Now, consider: $\dfrac{2}{x+4} - \dfrac{5}{x+2}$. At first sight, this may not seem like it has any-

thing to do with the previous example; however, it is still the subtraction of fractions, just algebraic fractions, as there is a variable. We will now examine how to complete this problem and how it relates to a basic subtraction of fractions problem. If this next part is too advanced for you right now, feel free to move on to the next section. You can always come back here! Otherwise, let us continue:

Simplify: $\dfrac{2}{x+4} - \dfrac{5}{x+2}$.

- We have the same issue as with the earlier arithmetic problem. We need to subtract, but we cannot, because the denominators are different. So, let us obtain a least common denominator.
- Since the denominators are $(x+4)$ and $(x+2)$, the least common denominator (LCD) must be the lowest quantity that both $(x+4)$ and $(x+2)$ can divide into. Think about this. Both $(x+4)$ and $(x+2)$ can divide

into $(x+4)(x+2)$. Watch this! $\dfrac{(x+4)(x+2)}{(x+4)} = x+2$. This is because $\dfrac{x+4}{x+4}$

cancels out, and $\dfrac{(x+4)(x+2)}{(x+2)} = x+4$ because $\dfrac{x+2}{x+2}$ cancels out.

- Therefore, the LCD will be $(x+4)(x+2)$. Let us set up the equivalent fractions:

- $\dfrac{2}{x+4} - \dfrac{5}{x+2}$

$\dfrac{}{(x+4)(x+2)} - \dfrac{}{(x+4)(x+2)}$. Now, like before, let us divide the old denominator into the LCD and multiply that product by the old numerator.

Left fraction: When we divide $(x+4)$ into $(x+4)(x+2)$, we are left with $(x+2)$. We then multiply $2(x+2)$. So, the numerator of the left fraction is $2(x+2)$.

Right fraction. When we divide $(x+2)$ into $(x+4)(x+2)$, we are left with $(x+4)$. We then multiply $(x+4)$ by 5. So, the numerator of the right fraction is $5(x+4)$.

- $\dfrac{2(x+2)}{(x+4)(x+2)} - \dfrac{5(x+4)}{(x+4)(x+2)}$

- In arithmetic, we simply needed to subtract the numerators, but now, we need to distribute. Keep in mind, we treat the 5 as a -5 since there is a negative sign in front of the 5.

$\dfrac{2x+4-5x-20}{(x+4)(x+2)}$. Now, simplify by combining like terms to get our final answer:

- $\dfrac{-3x-16}{(x+4)(x+2)}$.

It takes longer to add or subtract algebraic fractions; however, the goal and overall concept is the same as adding and subtracting fractions in arithmetic. Additionally, a scientific calculator would not be able to perform this operation. This is why you need to understand the basics of addition and subtraction of fractions, especially for college algebra and beyond.

A Look at Multiplication and Division of Fractions

Just like addition and subtraction, students need to have a conceptual understanding of multiplying and dividing fractions for more complex math. Consider a multiplication of fractions problem:

First Example: $\left(\dfrac{2}{5}\right)\left(\dfrac{5}{8}\right)$. The rule or algorithm for multiplication of fractions is to multiply the numerators and then multiply the denominators and ensure the final fraction is reduced. However, let us examine why. Consider an easier problem such as $\dfrac{1}{2}$ of 12. If there are 12 students in a room and half live on campus; then, there are 6 students who live on campus. However, when we take a fraction of a number, that is the same as multiplying $\left(\dfrac{1}{2}\right)(12)$. We can express 12, or any whole number, over 1 as $\dfrac{12}{1}$. So, $\left(\dfrac{1}{2}\right)\left(\dfrac{12}{1}\right)$, which is $\dfrac{12}{2}$, which is 6.

Let us return to the original example. There are two ways to approach this:

- We can multiply across $\left(\dfrac{2}{5}\right)\left(\dfrac{5}{8}\right)$, which is $\dfrac{10}{40}$. This reduces to $\dfrac{1}{4}$.
- We can cross reduce. Look at the cross values, 2 and 8 and then 5 and 5. Is there a common factor for 2 and 8? Yes, that is 2, and 2 divides into 2 one time. Also, 2 divides into 8 four times. How about 5 and 5? Yes, 5 divides into 5 one time. So, we can rewrite these as reduced fractions:

$\left(\dfrac{1}{1}\right)\left(\dfrac{1}{4}\right)$, which again is $\dfrac{1}{4}$.

Second Example: Now, let us look at division of fractions: $\dfrac{1}{2} \div \dfrac{1}{4}$. Derek gave us the rules for division of fractions earlier in this chapter. We keep the first fraction, change the operation to multiplication and flip (or invert) the second fraction and apply the rules of multiplication. There is a common saying attached to division of fractions, "Ours is not to reason why, just invert and multiply." In other words, just follow the rules. So here is one explanation of why this works.

Keep in mind that $\dfrac{1}{2} \div \dfrac{1}{4}$ can be expressed as $\dfrac{\frac{1}{2}}{\frac{1}{4}}$. This is known as a complex fraction. However, let us say we wanted to simplify the complex fraction by removing the bottom fraction. Since the bottom fraction has a denominator of 4, we need to multiply that fraction by 4 to cancel it out.

$$\frac{\frac{1}{2}}{\frac{1}{4} \cdot 4}$$

This cancels out the denominator because $\frac{1}{4} \cdot 4 = 1$. But wait! If we multiply 4 to the denominator, we must multiply 4 to the numerator. So,

$$\frac{\frac{1}{2} \cdot 4}{\frac{1}{4} \cdot 4}$$

Therefore, we wind up multiplying $\frac{1}{2} \cdot 4$, which is $\frac{4}{2}$, which simplifies to 2. But the rule of dividing fractions is to leave the first fraction, change the operation to multiplication, and invert the second fraction. Isn't that what we just did from the original division problem of $\frac{1}{2} \div \frac{1}{4}$? We kept the $\frac{1}{2}$, changed from division to multiplication, and inverted $\frac{1}{4}$ to 4. Therefore, the leave, change, invert method is a short cut!

How Does This Apply to Higher-Level Math?

Now consider a multiplication problem with algebraic fractions: $\frac{4}{x-2} \cdot \frac{x-2}{8}$.
We are going to apply the multiplication rule of fractions to this problem.

- We could multiply across and get: $\frac{4(x-2)}{8(x-2)}$. We would then divide 4 into

  8 and divide $(x-2)$ into $(x-2)$. This leaves us with: $\frac{1}{2}$.

- Or we could cross reduce. We would divide $(x-2)$ into $(x-2)$ and determine the greatest common factor between 4 and 8, which is 4. We would

  then divide 4 into both 4 and 8. This would give us: $\frac{1}{1} \cdot \frac{1}{2}$. Multiplying

  across gives us: $\frac{1}{2}$.

  Let us consider a division problem with algebraic fractions: $\frac{3x}{x+4} \div \frac{6x}{x+5}$.

- Let us first apply the initial rule of dividing fractions, which is to leave the first fraction, change the operation to multiplication, and

  invert the second fraction: $\frac{3x}{x+4} \cdot \frac{x+5}{6x}$.

- It may be simpler to try and cross reduce here. What is the greatest common factor between $3x$ and $6x$? $3x$ would divide into both terms. When looking at $(x+4)$ and $(x+5)$, there is no common factor. After cross reducing, we get: $\dfrac{1}{x+4} \cdot \dfrac{x+5}{2}$. Multiplying across, we get: $\dfrac{x+5}{2(x+4)}$.

  And no, we cannot cancel the x in the numerator with the x in the denominator! That is explained in Chapter 7's "Why in the World?"

Where Do Students Get Stuck with Fractions?

- Most of all, students mix up the rules (e.g., mistakenly using the multiplication rule when adding or subtracting fractions).
- Students struggle with the concepts of least common multiple and greatest common factor. More specifically, students may claim that the greatest common factor for 4 and 8 is 8. Additionally, students may claim that the least common multiple for 2 and 4 is 2.
- Because students do not understand the basic concepts of fractions with arithmetic, they do not know how to contend with algebraic fractions in higher-level math classes.
- A very common mistake in division of fractions is to cross reduce in division before changing to multiplication. Here is an example: $\dfrac{3}{10} \div \dfrac{5}{9}$. It looks tempting to try and cross reduce the 3 and 9 and the 5 and 10, but we cannot do that until we change the operation to multiplication!

Understanding the common errors can help you attack fractions. Take the approach that these common errors will not happen to you!

The Laws of Exponents

Even if you have never studied the Laws of Exponents, this topic just sounds terrible; doesn't it? Exponents sound complicated and boring, and now we must memorize "laws" that pertain to them?

What is an Exponent?

An exponent expresses repeated multiplication. It is a concept most students learn in later elementary or early middle school. For example, 2^4 indicates that we multiply 2 four times as in $(2)(2)(2)(2)$, which is 16. The exponent, which is known as the power, indicates how many times we multiply the base. Here are three other examples:

- $5^2 = (5)(5) = 25$
- $4^3 = (4)(4)(4) = 64$
- $3^3 = (3)(3)(3) = 27$

Certainly, some students struggle to understand the basics of exponents; however, this is generally not where the major struggles with exponents start.

Why do Students Hate Exponents?

> *I got exponents when I was in middle school, but when we started using exponents with x's, like x^4 times x^5, I just didn't get it.*
>
> George

> *Exponents started off simple, but like so many other things, when we started seeing x's attached to them, I just got lost.*
>
> Alicia

We can say this about so many mathematical concepts; they grow, and they develop.

Mathematical concepts start off simple and they become more complex, and exponents are no exception. Even when students understand the basics of exponents in arithmetic, exponents begin to look different in algebra, and students become frustrated, confused, and anxious.

What are the Laws of Exponents?

In algebra, there are three basic Laws of Exponents, and these involve the combination of variables and exponents. They are presented in Table 5.1.

TABLE 5.1

Laws of exponents

Law	Example	What it means	Answer
Product rule	$x^2 x^4 x^5$	x^2 times x^4 times x^5	We keep the base and add the exponents and get: x^{11}
Quotient rule	$\dfrac{x^8}{x^5}$	x^8 divided by x^5	We keep the base and subtract the exponents and get: x^3
Power to power rule	$(x^5)^2$	x^5 raised to the second power	We keep the base and multiply the powers and get: x^{10}

Wait? What?

So, I provide the rules, but this does not explain why we are doing this. For example, regarding the Product Rule, we multiplied exponents with 2, 4, and 5, and yet we added? That makes no sense. Let us examine each rule:

Product rule: We multiplied $x^2 x^4 x^5$, but what did we really do? x^2 means xx; x^4 means $xxxx$; x^5 means $xxxxx$. Let us put that all together, and we get: $xxxxxxxxxxx$. All together those are 11 x's, which is x^{11}. Of course, very few people expand exponents like I just did, so we use an algorithm, again known as the product rule, to add exponents with like bases when they are connected by multiplication.

Quotient rule: We divided $\dfrac{x^8}{x^5}$, but what did we really do? Expanding it out, like we did earlier, we get: $\dfrac{x^8 = xxxxxxxx}{x^5 = xxxxx}$.

But now let us cancel out the like terms. More specifically, we can cancel five x's in the numerator with five x's in the denominator: $\dfrac{xxxxxxxx}{xxxxx}$. That leaves us with xxx, which is x^3.

Therefore, the algorithm, for the quotient rule is to subtract the exponents with the like bases.

Power to power rule: We raised x^5 to the second power as $(x^5)^2$, but what did we really do? We took x^5 two times, which is $(xxxxx)(xxxxx)$. That is $xxxxxxxxxx$ or x^{10}. Therefore, the algorithm for raising a power to a power is to keep the base (x) and multiply the exponents.

Of Course it Gets More Complicated!

Where do students get stuck with exponents? They mix up the rules that we discussed. For example, students may multiply the exponents when they are supposed to add. Hopefully, the explanation of why we add, subtract, and multiply for each rule will help you master the rules. However, as problems with exponents become more complex, students may struggle more. Here is a more complicated example: Simplify: $\left(\dfrac{4x^4 y^3}{2x^2 y^2}\right)^3$. There are different ways to approach this problem. However, let us take it like this:

- Let us eliminate the parenthesis by raising everything inside the parenthesis to the outside (third) power. That means we raise each coefficient to the third power and apply the power-to-power rule for the exponents. We arrive at: $\dfrac{64x^{12} y^9}{8x^6 y^6}$.
- Since the remaining expression is attached by division, we will apply the quotient rule. We will divide the coefficients and subtract the powers (for the like bases) and arrive at: $8x^6 y^3$.

What About Negative Exponents? Do They Exist?

That sounds like a philosophical question, but yes, we see negative exponents throughout mathematics. Here is an example of a negative exponent: x^{-3}. The bottom line is we cannot leave an exponent as negative. Therefore, we make the exponent positive by using the reciprocal. Since we can assume x^{-3} is $\dfrac{x^{-3}}{1}$, the reciprocal is $\dfrac{1}{x^3}$.

Let us look at one reason why this happens. Consider the following problem: $\dfrac{x^5}{x^8}$. We need to apply the quotient rule, which means we subtract the exponents $(5 - 8)$ and get x^{-3}. However, let us take a closer look and expand the exponents as we did earlier: $\dfrac{xxxxx}{xxxxxxxx}$. Let us cancel out the like x's: $\dfrac{xxxxx}{xxxxxxxx}$. This leaves us with $\dfrac{1}{x^3}$. Again, when presented with a negative exponent, we can simply make the exponent positive by using the reciprocal.

Where Will I See the Laws of Exponents?

The laws of exponents may be introduced as early as an introduction to algebra class, and the problems get more complicated as the algebra progresses. Keep in mind, however, that it all comes back to the three basic laws of exponents that we discussed. The laws of exponents continue to be applied in college algebra, calculus, and beyond.

Exponents can be Weird

There are two very strange concepts of exponents. We will tackle one of those strange concepts in "Why in the World" at the end of the chapter. Right now, let us address what is the difference between $(-3)^2$ and -3^2? Isn't the answer to both expressions 9 since we multiply $(-3)(-3)$? Not exactly. Let us first examine $(-3)^2$. What does this mean? This indicates $(-3)(-3)$, which is 9. But what about -3^2? The exponent binds directly to the base, which is 3. Therefore, we can view -3^2 as $-(3)^2$, which is -9. Another way to view -3^2 is the negative of 3^2.

Probability

Probability is how likely an event or outcome is to happen. As the simplest example, if we flip a two-sided coin, we may ask, "How likely is it that the coin lands on heads?" We are asking "What is the probability that the coin

lands on heads?" Probability can be expressed as a fraction, which consequently has a numerator and a denominator. The numerator reflects the
number of favorable outcomes, which is "1" since there is only one head. The
denominator indicates the total possible number of outcomes, and since there
are two possible outcomes in flipping a coin (heads or tails), the denominator
of the fraction is 2. Therefore, the probability of obtaining heads is $\frac{1}{2}$ or 50%.
That seems simple, right? So why is probability despised?

Why is Probability Hated?

> *I always hated probability because there are so many fractions. It's just fraction,
> after fraction, after fraction. I hate fractions, so yeah, probability always made
> me miserable.*
>
> Joanne

> *I think it was just so much crazy wording in probability that always made me feel
> so lost. Like "with replacement" and "without replacement." I never understood
> what the heck we were replacing.*
>
> Denise

> *It's just so many words. It's all word problems, and they were also so hard. I
> mean, if you asked me what's the probability of rolling a 4 on a die, I could answer
> that, but probability always had these hard word problems.*
>
> George

> *I just remember the time period was terrible. Probability came up in middle
> school, and that was when math just got really hard, and I just hated middle
> school.*
>
> Alicia

Students detest probability for a variety of reasons. Joanne made an excellent point. As discussed in the previous section, probability, in its most basic
form, is a fraction. Consequently, the more complex the problem is, the more
fractions will be involved.

I also want to share a personal anecdote with my teaching of probability. In the fall of 2018, I was assigned a quantitative reasoning class,
which included probability. It had been several years since I taught probability, and I noticed that my students were struggling. I asked them to
tell me what their barriers were in learning the material. Several students
discussed how the wording and semantics of probability applications can
be confusing. Denise provided a good example in the above testimonial.
Therefore, I made it a priority to ensure that, as a class, we were focusing

on the confusing terminology of a probability application and providing clarity.

Alicia also made a good point, and we will see this later with another despised topic. Upper elementary school and middle school mathematics can be difficult for many students. This is where concepts become more rigorous and abstract, and algorithms become longer. Students must apply basic information to more complex concepts, and if students have gaps with the basic information, this can be an ingredient for failure. More specifically, students learn about basic probability in elementary school, and they learn about fractions. However, in middle school, they must apply these basic concepts of probability, fractions, and word problems to more complex probability.

When Will I Need Probability for College Math?

Students who need to complete either quantitative reasoning or introduction to statistics will need to study probability. As I mentioned in Chapter 3, introduction to statistics consists of a deep study of probability. Students who need to complete a finite mathematics class, either for business calculus or linear algebra, will also need to master some probability. Probability is also covered in the teacher preparatory math classes. College algebra generally does not cover probability; however, students may see various forms of probability when completing their calculus sequence.

Do I Need to Prepare for Probability in Advance?

Most college classes that utilize probability will start with the basic concepts of probability. However, the topic will progress quickly; therefore, like every other mathematical topic, it is important to get help when something becomes difficult. Additionally, when studying probability, classes utilize either the scientific or graphing calculator. Therefore, it would be a good idea to become familiar with basic operations (e.g., operations with fractions) on these calculators. We will explore probability further in Chapter 6 since probability generally consists of word problems. Consequently, we will explore how to approach basic probability and where students can get "stuck."

Geometry: Angles, Lines, and Rays...Oh My!

Geometry is the study of shapes, angles, objects, and dimensions. It is a large branch of mathematics. This is why students start studying geometry in elementary school, continue through middle school, and spend a year studying the subject in high school. Consequently, students have brought up geometry as a subject that has brought them difficulty. Specifically, students have identified the study of angles, which includes lines and rays, as a problematic area.

Why is This Topic so Bad?

> *I just remember never really understanding what an angle was when I was in elementary school, so everything about angles after that never made sense.*
>
> Emily

> *Geometry started off simply enough when it was like find the perimeter or area of a rectangle or find the circumference of a circle, but when it got to angles, rays, and those arcs in a circle, I was just lost.*
>
> Peter

> *Just the wording used to confuse me: rays, vertex, acute angles, interior, exterior angles. I just remember being totally lost.*
>
> Joanne

> *Complimentary and supplementary angles, that was so confusing. Then there was stuff about side-angle-side, and angle-side-angle. That was really confusing. I remember on tests, when we were supposed to prove stuff, I would just guess and write "side-angle-side," or "side-side-side" because I didn't know what I was doing. Of course, I got no credit.*
>
> David

Like probability, a common concern, regarding angles, lines, and rays, is the complex terminology. As I discussed in Chapter 4, math is filled with terminology, but this is especially the case for geometry. Moreover, when students do not understand the meaning behind angles, rays, vertices, and various types of angles, they will fall behind very quickly. This is a challenge because, like fractions, students learn these complex concepts when they are still at the concrete operational stage. Quite frankly, it can be difficult for young children to understand intricate concepts such as angles, rays, planes, etc. By middle school and high school, students have transitioned into what Jean Piaget labeled the formal operational stage, where they can understand more abstract concepts; however, they likely have too many gaps in their geometry knowledge base by then or are simply too far behind in geometry.

Like other mathematical concepts, geometry becomes more complex in the upper elementary school/middle school period. Additionally, during this time, students are introduced to algebra and must master concepts such as set theory and more complex word problem applications. The upper elementary/

middle school period can be overwhelming in terms of math concepts, but this time can be difficult for children in general. Hormones are changing, and children may contend with issues such as anxiety and depression.

Where is Geometry Used in College-Level Math Classes?

Again, "geometry" includes a great deal of material. Concepts such as the perimeter of a rectangle, circumference and area of a circle, and area of a triangle are often used in developmental math classes. However, there is likely little to no geometry in quantitative reasoning or statistics. What about the dreaded concepts of angles, lines, rays, etc.?

Students who need to complete the teacher preparatory classes will likely see angles, lines, rays, and planes. These students will also need to master some geometric proofs using congruent triangles and other theorems.

Students who will need to complete trigonometry and a traditional calculus class must master the basics of angles, lines, rays, vertices, and congruent angles. This is imperative to completing trigonometry and for several applications in calculus. Additionally, in calculus, students will need to master conic sections, which includes the study of circles, cones, cylinders, and spheres.

Do I Need to Prepare for Angles, Lines, Rays, etc.?

This is hard to answer. Teacher preparatory classes may begin with the basics of angles, lines, and rays, but it would be helpful to view some videos on these concepts. Students who need to complete trigonometry need an incoming understanding of angles, lines, rays, alternate interior angles, alternate exterior angles, and congruent angles. See below for a brief description of some of the basic geometric terms discussed in this section.

- Line: Figure 5.1 shows an example of a line, which is a set of collinear points that extend infinitely in each direction.

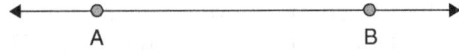

FIGURE 5.1
A line

- Ray: Figure 5.2 shows an example of a ray, which is a line with one end point that extends infinitely in one direction.

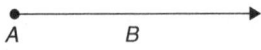

FIGURE 5.2
A ray

- Angle: Figure 5.3 shows an example of an angle, which is a geometric figure formed when two rays meet. Angles are measured in terms of degrees.

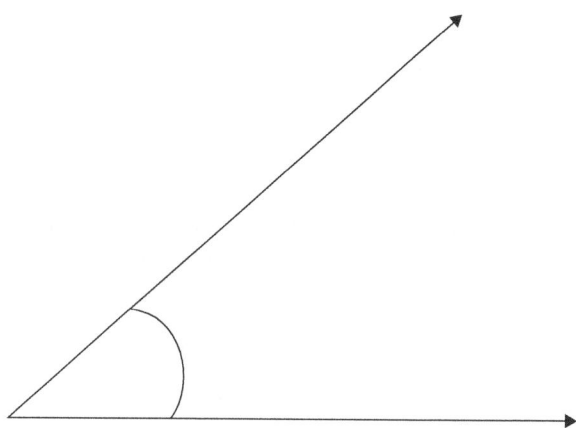

FIGURE 5.3
An angle

- Vertex: Figure 5.4 shows an example of a vertex, which is the point where two rays meet.

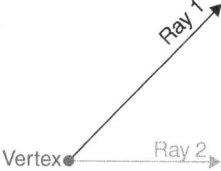

FIGURE 5.4
Vertex

- See Figure 5.5 for an example of two complimentary angles, which are angles that, when added together, equal 90 degrees. We label the 30-degree angle (∠) as ∠AOD. We label the 60-degree angle as ∠DOB. Additionally, it would not make sense to label either angle, ∠O. Why not? Because both angles share a common vertex.

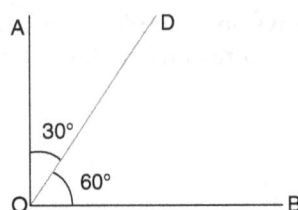

FIGURE 5.5
Complimentary angles

- Supplementary angles: Figure 5.6 shows an example of two supplementary angles, which are two angles that, when added together, equal 180 degrees.

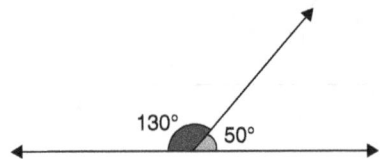

FIGURE 5.6
Supplementary angles

- Congruent angles: Figure 5.7 shows an example of two congruent angles. These are two angles that have the same measurement. While ∠CBA and ∠RQP are two different angles, they both measure 40 degrees, making them congruent angles.

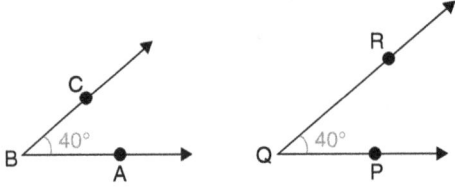

FIGURE 5.7
Congruent angles

- Complexity added to angles: Consider Figure 5.8.

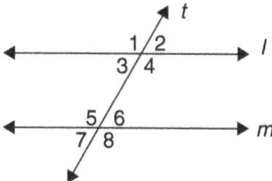

FIGURE 5.8
Complex angles

Let us discuss what we are seeing. There are two lines, labeled l and m. Lines l and m are parallel lines in that they are an equal distance from each other but will never meet. Lines l and m are cut by line t, which is a transversal: a line that intersects two or more lines. There are eight angles in this diagram. Since a transversal has intersected two parallel lines, here is what we know:

- ∠1 and 2, ∠3 and 4, ∠5 and 6, ∠7 and 8 are supplementary angles. How do we know these pairs of angles add to 180 degrees? Because a straight line always measures 180 degrees. For example, ∠1 and ∠2 are two angles on a straight line; therefore, they are supplementary. Additionally, ∠1 and ∠3, ∠2 and ∠4, ∠5 and ∠7, and ∠6 and ∠8 are also supplementary as those pairs of angles add to 180.

- Which angles are congruent? More specifically, which pairs of angles have equal measure? For starters, ∠1 and ∠4, ∠2 and ∠3, ∠5 and ∠8, and ∠6 and ∠7 are congruent angles in that each pair has the same measurement. How do we know this? Because these are opposite angles, and opposite angles are congruent. We can also say, as an example, < 1 ≅< 4. That symbol between the two angles indicates congruency.

- ∠1 and ∠8, ∠2 and ∠7 are congruent angles. How do we know this? Because these are alternate exterior angles. Look at ∠1 and ∠8, for example. These angles are on opposite sides (alternate) of the transversal and outside (exterior) of lines l and m.

- ∠4 and ∠5, ∠3 and ∠6 are congruent angles. How do we know this? Because these are alternate interior angles. Look at ∠4 and ∠5, for example. These angles are also on opposite sides (alternate) of the transversal, but they are inside (interior) lines l and m.

- We also know that, and let us use the symbol this time, < 1 ≅< 5, < 2 ≅< 6, < 3 ≅< 7, < 4 ≅< 8. How so? These are corresponding angles. When a transversal crosses a pair of parallel lines, pairs of angles are formed on the same side of the transversal and in the same relative position. Those are corresponding angles.

Factoring

Factoring is writing a product (the answer to a multiplication problem) in its simplest possible factors. Consider the number, 6. The simplest way to write the factors (numbers that multiply to a product) is $2 \cdot 3$. How about 8? Yes, $2 \cdot 4 = 8$. However, we can simplify this, because $2 \cdot 2 = 4$. Therefore, 8 in factored form is $2 \cdot 2 \cdot 2$. How about 12? Yes, $4 \cdot 3 = 12$; however, $2 \cdot 2 = 4$. Therefore, 12 in factored form is $2 \times 2 \times 3$.

Of course, the examples above are from basic arithmetic, and the type of factoring that gives students difficulty is in algebra, which we will examine later.

Why do Students Hate Factoring?

I could never do math backwards, things like what 2 numbers multiply to 18 but add to 3.

Jessica

Too much of what multiplies to this number and subtracts to that number and then all signed numbers. It's like I can add or multiply, but don't start mixing them together and make it all confusing.

Sam

I had a professor of math education who once told me that "factoring is all about numbers." She was correct. To understand factoring, a student needs a strong number sense. Students must know their addition, subtraction, and multiplication facts and have strong divisibility skills and be able to perform mental math. Additionally, students must have a thorough understanding of signed numbers. Students who do not have a strong number sense will struggle with factoring.

Why Does Factoring Exist and When Will I Need it?

Any student who has struggled through factoring in an algebra class has asked this question in frustration. Is factoring just something someone invented to torture students? Not exactly. Factoring is used to simplify larger problems and applications. When solving mathematical problems, the solutions must be in simplest form. We must use certain methods to get those answers into simplest form, and factoring is a way to simplify. This can vary between colleges, but unless you are a STEM major, you may not encounter factoring, although you may see some basic factoring in introductory algebra classes. Factoring, however, is not generally covered in quantitative reasoning or statistics. However, you can bet on seeing factoring in intermediate algebra, college algebra, trigonometry, and beyond!

So, Let's Factor!

As mentioned earlier, factoring is about number sense. There are several types of factoring. However, in this section, we will examine two basic types of factoring to get you started.

First Type of Factoring

This type of factoring is entitled, factoring out the greatest common factor (GCF). To get started, review the questions and answers in Figure 5.9.

Let us bring some algebra into this. How would you multiply $12(x+3)$? We would distribute (multiply) the 12 to both the x and the 3. Our solution is $12x+36$. However, let us say we wanted to go the other way. We wanted to determine what was multiplied to obtain $12x+36$. More specifically, we want to factor $12x+36$.

- First, we look for the GCF for the two terms. Below are the factors for both 12 and 36:

  $12 = 1, 2, 3, 4, 6, 12$

  $36 = 1, 2, 3, 4, 6, 9, 12, 18, 36.$

  We can see that 12 is the GCF for 12 and 36.

- We pull out a 12 and divide 12 into both 12 and 36, and that is how we obtain $12(x+3)$.

  How about another example? Let us factor $9x+15x^2$. More specifically, we want to obtain what was multiplied to obtain $9x+15x^2$.

- What is the greatest number (GCF) that can divide into both 12 and 18? The answer is 6.
- What is the greatest number (GCF) that can divide into both 32 and 48? The answer is 16.
- What is the greatest number (GCF) that can divide into both 48 and 64? The answer is 16.

FIGURE 5.9
Review of greatest common factor.

- First, we look for the GCF for the two terms. Below are the factors for both 9 and 15.

$9 = 1, 3, 9$

$15 = 1, 3, 5, 15.$

We can see the GCF is 3, but we are not done. Notice how there are x's in both terms?

$x = x$

$x^2 = x \cdot x$

Therefore, there is a GCF of x as well. So, the GCF is $3x$.

- We pull out a $3x$ and divide it into both terms: $3x(3 + 5x)$. See how we applied the quotient rule when factoring? $\dfrac{9x}{3x} = 3$ and $\dfrac{15x^2}{3x} = 5x.$

Second Type of Factoring

In this section, we will examine factoring trinomials (three terms). Before we do that, it is important to understand, just like for the first type of factoring, how to multiply various terms. For example, can you multiply $(x - 3)(x + 5)$? We multiply (or distribute) the x to both terms in the second binomial, and we multiply the –3 by both terms in the second binomial. So, let us do that: $(x)(x) + 5(x) - 3(x) - 3(5)$. This gives us $x^2 + 5x - 3x - 15$. After combining like terms, we get: $x^2 + 2x - 15$.

As mentioned earlier, signed numbers are a major part of factoring, and students must know them literally backwards and forwards. So, before we get into factoring trinomials, try the exercises in Figure 5.10 and check your

- What are two numbers that multiply to 28, but those two numbers must also add or subtract to –3? The answer is – 7 and 4.
- What are two numbers that multiply to 42, but those two numbers must also add or subtract to 1. The answer is 7 and –6.
- What are two numbers that multiply to 36, but those two numbers must also add or subtract to –9. The answer is – 12 and 3.

FIGURE 5.10
Practice for factoring.

answers. Additionally, this is why many students struggle with factoring. They struggle with the prerequisite concepts in Figure 5.10.

Now, let us try and factor a trinomial. Specially, let us factor $x^2 - 7x + 12$. We want to determine what was multiplied to arrive at that trinomial. Since it is a trinomial, we know this will likely factor into two binomials, as shown earlier. So, we know our answer will look like this ( )( ). We just need to fill in the blanks! The easy part is the first term in each parenthesis. We know that $x \cdot x = x^2$, so we can fill in $(x\ \ )(x\ \ )$. How about the remaining two terms? What pair of numbers multiply to 12 (the third term of the trinomial) but also add to the middle coefficient, which is –7? How about –4 and –3? So, $x^2 - 7x + 12$ factors into $(x-4)(x-3)$. We can always check our work by multiplying $(x-4)(x-3)$ to see if that leads us to the original trinomial.

So, $(x)(x) - 3(x) - 4(x) - 4(-3)$. And yes, we get: $x^2 - 3x - 4x + 12$, which simplifies to $x^2 - 7x + 12$.

Let us factor one more trinomial: $x^2 - x - 20$. Again, since it is a trinomial, this will likely factor into two binomials. We also know that $x \cdot x = x^2$. Already, we know our answer looks like this: $(x\ \ )(x\ \ )$. What pair of numbers multiply to the third term, which is 20 (do not worry about the sign of the third term) and add to the middle coefficient, which is –1? That would be –5 and 4. So, $(x-5)(x+4)$. Does that check? $(x)(x) + 4(x) - 5(x) - 5(4)$. This gives us $x^2 + 4x - 5x - 20$, which simplifies to $x^2 - x - 20$.

Factoring can be overwhelming, and we just scratched the surface of factoring. As I mentioned earlier, there are other types of factoring and more complex examples. Let us review and summarize the major prerequisite skills for factoring, especially where students "get stuck."

- Possessing solid divisibility skills, especially being able to compute the greatest common factor (see Figure 5.9).
- Understanding all operations of signed numbers, forwards and backwards (see Figure 5.10).
- Understanding the basic laws of exponents, particularly the product and quotient rule.

Slope of a Line

Earlier in this chapter, we discussed the problems with angles, lines, and rays. Well, lines bring about another set of problems for students. The slope of a line, which may bring a flutter of nerves to your stomach, is a difficult topic for many students.

Before we discuss the slope of a line, we need to remember that a line is made up of two or more collinear points. It is imperative to be able to graph such points on a Cartesian plane, which is made up of an x-axis (horizontal) and a y-axis (vertical). See Figure 5.11 and try to plot the following points: $(-2,-2)$, $(0,0)$, $(1, 1)$. The answers are in Figure 5.12. Remember, the first coordinate refers to the x-axis; the second coordinate refers to the y-axis.

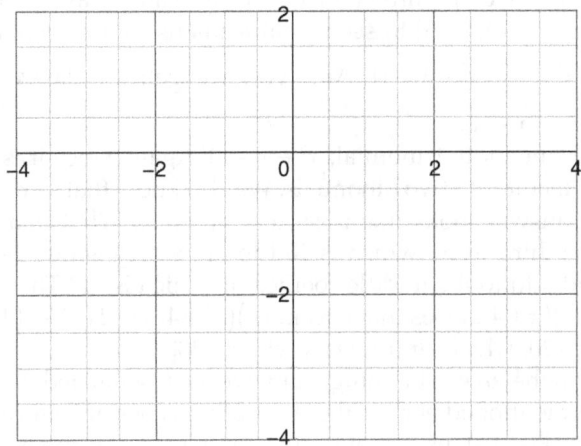

FIGURE 5.11
A Cartesian plane

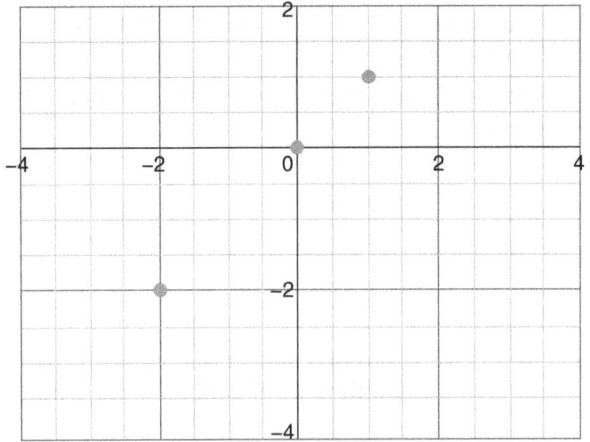

FIGURE 5.12
Points on a Cartesian plane

Students may need to review the concept of plotting points on a Cartesian plane before attempting to find the slope.

How did you do?

When attempting to comprehend slope, consider a hill or even a mountain where there is a slant. The slope is the rate of change. In fact, consider the line below in Figure 5.13. On the line, move from left to right. Additionally, when reading any graph, always move from left to right.

Start where the slanted line crosses the vertical line (the y-axis). That is at the point (0, 1) and move upward and to the right. Notice, how we move up three spaces and over two spaces (to the right) to arrive at our next point (2, 4)? Continue to move upward and to the right, and we move another three spaces upward and two to the right before the next point (4, 7) and this pattern will continue. Therefore, the slope of the line is $\frac{3}{2}$. This is the rate of change, but it is also a constant rate of change. As this line continues, the line will increase by three and run over by two. Additionally, lines have names. For example, the line in Figure 5.13 is $y = \frac{3}{2}x + 1$. The slope is $\frac{3}{2}$, but the line crosses the y-axis at 1. This is also known as the equation of the line.

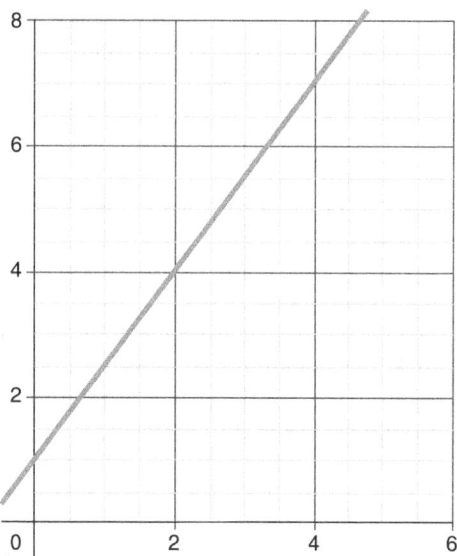

FIGURE 5.13
A line on a Cartesian plane

Where is Slope Used?

Slope represents the rate of change. An example in real life is the depreciation of an item. A car or a mobile home loses value over time. If the average loss is the same each year, we use slope to determine this rate of change. Designing a bridge also utilizes slope. A bridge will increase gradually to its peak, and that constant gradual increase is the slope.

Where will you see slope in your mathematical studies? Most developmental and college-level math classes utilize slope in some form. For STEM students, mastery of slope is a must. In fact, to quote one of my former professors, "They won't let you in the front door of calculus if you don't understand slope."

Why do Students Hate Slope?

> *It's just confusing and boring. Everything about slope and equation of a line is just confusing. Just the words used to lose me.*
>
> Derek

> *Slope and all that stuff is so progressive. If you don't get one thing, you'll be lost forever. I remember I was struggling with just putting a point on a graph and understanding that. Then, we went to slope, equation of a line, and all these formulas, and I was just hopelessly lost.*
>
> Joyce

> *I hated slope in algebra, but I'll admit, it made more sense when I could see how slope was applied in real-life. It wasn't just a bunch of boring formulas to memorize.*
>
> Joanne

That sums it up. Slope is boring, confusing, and progressive to the point where the lack of understanding of any concept leads to confusion going forward. Joanne also makes an interesting point. Seeing how slope is applied to real life can make it more interesting and less daunting. This is one example of where the corequisite model that combines developmental material with a college-level class, such as quantitative reasoning, can be more advantageous than a traditional algebra class. In the corequisite model, students can learn about slope but immediately apply it to real–life.

Slope has a Formula

We discussed how we can use a graph to determine the slope (or rate of change of a line), but what if we did not have a graph on hand or did not want to create a graph? Here are two points: $(-3, 3)$ and $(0, 2)$. We can determine the

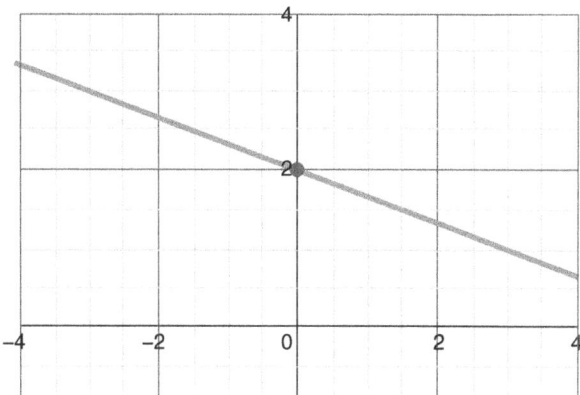

FIGURE 5.14
A line from two points

slope by using the following formula: $\dfrac{y_2 - y_1}{x_2 - x_1}$. That does not make sense; does it? Keep in mind that the first coordinate of a point is the x-coordinate, and the second coordinate is the y-coordinate. Therefore, –3 is the x-coordinate, and 3 is the y-coordinate. But what about the second point? How about if we label –3 as x_1 and 3 as y_1, since there are two x's and two y's. Therefore, we can label 0 as x_2 and 2 as y_2. Let us plug everything into the formula: $\dfrac{2-3}{0-(-3)}$. This simplifies to $-\dfrac{1}{3}$. How about we graph the points with a line to see if we are correct?

Notice how if we start at the point $(-3, 3)$ and travel down (negative) one coordinate and three coordinates to the right (positive), we arrive at the next point? Additionally, the formula did not come from outer space. The slope formula is the difference in the y-coordinates divided by the difference in the x-coordinates. When we travel from one point to the next, we go up or down (y-axis) and then left or right (x-axis). Finally, we can label the line in Figure 5.14, $y = -\dfrac{1}{3}x + 2$, because the slope is $-\dfrac{1}{3}$ and the y-intercept is 2.

Another Way to Determine Slope

As mentioned earlier, slope is a rate of change. Let us say that Randy is on the football team and trying to gain weight each week to prepare for the season. Table 5.2 shows his weight for each week.

We can see from the table that Randy's weight has increased by three pounds each week. "Week 0" indicates his starting weight. "Week 1" is after one week's time. Since Randy's change in weight is constant, we can

TABLE 5.2

Determining slope

Week	Randy's weight
0	160
1	163
2	166
3	169

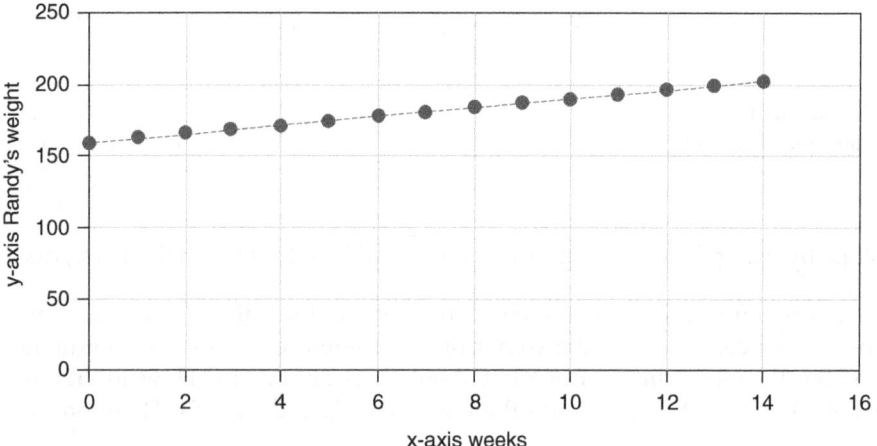

FIGURE 5.15

Graph of Randy's weight

determine an equation of the line, which would be $y = 3x + 160$. The slope is 3, and the y-intercept is 160. That is because 160 is the starting point. We could use this equation of the line to predict future values. For example, what would Randy's weight be after six weeks? Keep in mind that the x-coordinates represent the weeks, and the y-coordinates represent Randy's weight. See Figure 5.15 for the graph of this situation. Therefore, we will insert a "6" for x and determine the y-value: $y = 3(6) + 160$. This simplifies to 178. Consequently, if this trend continues, Randy will weigh 178 pounds after six weeks. How about if we wanted to determine how long it will take Randy to get to 200 pounds? In this case, we are solving for "x," as this variable refers to time, and we know "y" is 200. So:

$200 = 3x + 160$. We need to get the x by itself, so we will subtract the 160 from both sides:

$$200 = 3x + 160.$$

$$-160 \qquad -160$$

$40 = 3x$. Now, divide both sides by 3 to get the x by itself.

$$\frac{40}{3} = \frac{3}{3}x$$

Therefore, $x = \frac{40}{3}$ or 13.3 weeks. This means that it will take a little more than 13 weeks for Randy to hit 200 pounds.

Should I Prepare for Slope?

If you place into a developmental math class or a college-level class with a booster component, the instructor will likely introduce slope of a line from the beginning. However, as I mentioned earlier, students are expected to understand slope in higher-level classes such as pre-calculus or calculus. Of course, it never hurts to practice the basic concepts, which we will do very soon!

Why in the World? Part 5

Why in the World is Anything Raised to the Zero Power Always 1?

For example, $5^0 = 1; 8^0 = 1; x^0 = 1$.

Why this makes no sense? How can we raise anything to the zero power and get 1? Do I need to say more?

Let us try to understand: To best explain let us utilize a base of 10. Would you agree with the following? $10^4 = 10,000; 10^3 = 1,000; 10^2 = 100; 10^1 = 10$. Looking at the solutions, do you see a pattern? They are in descending order for place value: ten thousands, thousands, hundreds, and tens. What place is next? The one's place. So, $10^0 = 1$. How about 10^{-1}? That is $\frac{1}{10}$, which is 0.1 or one-tenth. How about 10^{-2}? That is $\frac{1}{10^2}$ or $\frac{1}{100}$ or 0.01 or one-hundredth. Therefore, the zero power is just a place holder between the whole values and the decimals.

Here is another example with a different base and another explanation. See if you can identify the pattern:

$$2^3 = 8; \ 2^2 = 4; 2^1 = 2; 2^0 = 1; 2^{-1} = \frac{1}{2}; 2^{-2} = \frac{1}{4}.$$

Look at the solutions. Start with the first answer (8) and divide by the base (2), and you will arrive at the next answer (4)! As you move to the left, it continues to work: $4 \div 2 = 2; 2 \div 2 = 1$, and that covers the zero exponent. Keep moving to the left, and this establishes a pattern for the negative exponents as well: $1 \div 2 = \dfrac{1}{2}; \dfrac{1}{2} \div 2 = \dfrac{1}{4}$.

Practice Problems

Hopefully, this chapter has helped you feel more at ease with some difficult topics. More so, after reading the student testimonials you realize that you are not alone if you have struggled with these topics. Below, I have provided some practice problems for each topic. The answers are in Appendix C. There may be some overlap with the practice problems in Chapter 3, which served as sample prerequisites for various classes. However, there is no such thing as too much practice in math! Again, we will examine probability more closely in Chapter 6.

Practice with Fractions. Be sure to Reduce the Fractions

1) Add: $-\dfrac{3}{5} + \dfrac{2}{3}$

2) Subtract: $\dfrac{9}{10} - \dfrac{2}{3}$

3) Multiply: $\left(-\dfrac{3}{7}\right)\left(-\dfrac{14}{21}\right)$

4) Divide: $\left(-\dfrac{4}{5}\right) \div \left(\dfrac{12}{15}\right)$

   Go ahead and try these!

5) Add: $\dfrac{3x}{x-2} + \dfrac{2}{x+5}$

6) Multiply: $\dfrac{4}{5x} \cdot \dfrac{15x}{12}$

7) Divide: $\dfrac{x-3}{5x} \div \dfrac{x-3}{10}$

Practice with exponents. Simplify the following:

1) $x^5 x^{10} x^{11}$

2) $\dfrac{y^{12}}{y^4}$

3) $(y^{10})^4$

4) $(\dfrac{4x^5 y^4}{2x^3 y^4})^2$

5) $(\dfrac{3y^4 z^5}{6y^5 z^3})^2$

6) Rewrite as a positive exponent: x^{-6}

7) Rewrite as a positive exponent: y^{-2}

Geometry: Angles, Lines, and Rays

1) Is the following a ray or a line? Why?

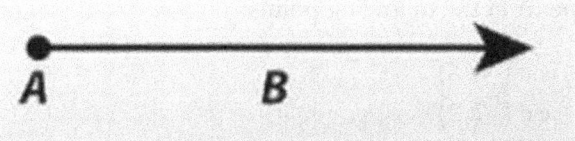

2) Consider the diagram below:

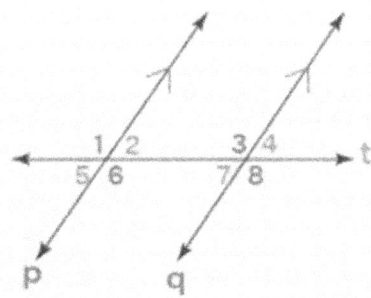

Lines p and q are cut by the transversal, t.

a) Identify four pairs of angles that are supplementary.
b) Identify the opposite angles.
c) Identify the alternate interior angles.
d) Identify the alternate exterior angles.
e) Draw an example of two congruent angles.

Factoring

Factor the following completely:

1) $5x^2 - 15x$
2) $9x^2 - 21x$
3) $10y^2 - 30y^3$
4) $x^2 - 4x - 45$
5) $x^2 + x - 72$
6) $x^2 - 5x - 84$

Slope

Find the slope from the following points:

1) $(-4, 5)$ and $(-6, 8)$
2) $(-3, 6)$ and $(-7, 2)$
3) $(2, 5)$ and $(-4, 7)$
4) Using the below graph, identify the equation of the line.

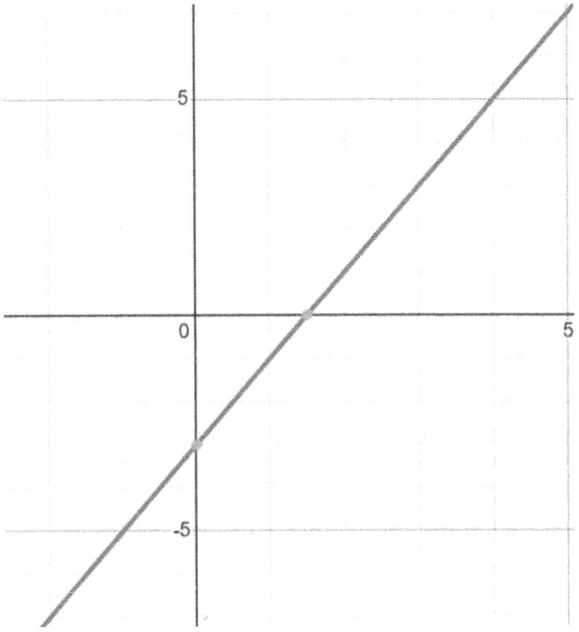

Activity

- In this chapter, we have discussed several math topics in which students struggle, and in the next chapter we will discuss word problem applications. Have you had trouble with these topics? Do you agree with the student statements?

- Can you think of additional math topics in which you have struggled? What are they? Write about some of your experiences as to why such topics were so troublesome.

- Do any of the tips and suggestions for getting started with the topics in this chapter help or make you feel less anxious? If so, how?

- Regarding the math topics that give you concern, what can you do to be proactive going into your math class. More specifically, what can you do to set yourself up for success?

6

Word Problems: The Bane of My Existence

In Chapter 5, we examined six mathematical topics that create difficulty for students. However, the most troublesome topic gets its own chapter, and that topic is word problems! Therefore, in this chapter, we will unpack the struggle with word problems. Again, keep in mind that some of the examples of word problems that we will examine may be difficult for you to understand at this point, because you may not have mastered the prerequisite skills, but my goal is to give you strategies to attack word problems.

Why do Word Problems Bring Out the Anxiety?

Picture this. You are completing a homework assignment or even an exam and you encounter the following:

> A gardener is planting two types of flowers: tulips and daffodils, in a row. The number of tulips is 5 more than twice the number of daffodils. If there are 35 flowers in total, how many tulips are there?

When does your anxiety set in? Is it after reading the problem once? Is it after reading one sentence? Or does just the sight of this problem send you into a panic? We will attack the above problem later in this chapter but let us first examine why word problems set off such anxiety.

Reading Comprehension or Reading Differences

Captain Obvious said that word problems involve reading, but is your reading comprehension at the level of the word problem? Not being able to properly read a word problem can be a major barrier and create a lot of anxiety and frustration.

> *I always hated math, but word problems were the worst. In high school and before that, I had such a hard time getting through any word problem, and I just*

DOI: 10.1201/9781003614142-6

thought I was stupid. People treated me like I was stupid. Turns out, I was just way behind on my reading. When I got to community college, I was able to get help with my reading in this adult education program. Once I read better, I was amazed how much better I could be at word problems.

Emily

Colleges used to offer various developmental reading classes to provide students with the tools to read at a college level. However, in recent years, many of these classes have been eliminated. The bottom line is if you are concerned about your reading level going into college, you will want to address that immediately. It is possible that the results of your placement exam may suggest that you need additional assistance with reading. This could be a developmental reading class or a study skills class. Feel free to ask your academic advisor if there is anything you can do to ensure that you are reading at or close to the college level.

It is also possible that you are one of the 780 million people in the world who have dyslexia. Whether you are aware that you have it or if you suspect that you have dyslexia, the disability services at your college can help. Addressing a reading challenge is not only important for working on math word problems but also for reading-intensive classes such as English, psychology, or sociology.

Not Even Knowing Where to Start

Consider the type of problem below that we addressed in Chapter 5:

Find the slope of a line through the following points: $(4, -5)$ and $(-3, 8)$.

Once students learn how to find the slope of a line, they will know where to start, which is the formula for slope. More specifically, there is a definitive first step of the process. That is not the case for word problems.

One thing I always hated about word problems is after I read them, I had no idea where to even begin, so I just started guessing and writing stuff down and going in the wrong direction.

Louis

I'll admit it. When I read a word problem, I just started writing things down. I guess it made me feel better to at least write something down rather than sit there and think about it. Does that make sense?

Nancy

Not knowing where to begin a mathematical application can certainly create a sense of anxiety and panic. Consequently, instead of thinking the problem through and devising a plan, many students, like Louis and Nancy, begin to guess and write anything down. I also think Nancy makes a great deal of sense. Sitting and thinking about a problem can create a lot of anxiety. Students have a fear of being uncertain, and consequently travel down any path, even the wrong one, to alleviate that anxiety. Any mathematical application needs a starting point. However, word problems may take more thought to get started. Later in this chapter, we will examine ways to start and work through a word problem.

Math Terminology is an Issue

In Chapter 4, we discussed the importance of math vocabulary. This is especially important when it comes to word problems. For example, in the word problem listed earlier, students would need to understand that "twice the number" means two times as much and that "more than" indicates addition, and it will get much more complicated than that. For example, terms such as "maximizing" or "minimizing" profits indicate that the application deals with linear programming, a concept in college algebra or finite mathematics. In probability, the terms "and" and "or" indicate a specific type of probability and a precise formula. See below (Tables 6.1–6.5) for some tips in getting started in translating basic word problems with the four basic operations (addition, subtraction, multiplication, and division). Moreover, this can get you started in an introductory or pre-algebra class.

Notice how in Table 6.1 we were able to switch the expressions around and still be correct? That is because the operation of addition is commutative.

TABLE 6.1

Addition Key Words and Phrases

English	Mathematics
The sum of 5 and a number	$5 + x$ or $x + 5$
6 more than a number	$6 + x$ or $x + 6$
The sum of a number and 12	$x + 12$ or $12 + x$
18 added to a number	$18 + x$ or $x + 18$

TABLE 6.2

Subtraction Key Words and Phrases

English	Mathematics
The difference between a number and 10	$x - 10$
12 less than a number	$x - 12$
A number less than 12	$12 - x$
21 subtracted from a number	$x - 21$
A number subtracted from 21	$21 - x$

TABLE 6.3

Multiplication Key Words and Phrases

English	Mathematics
Six times a number	$6x$ or $x6$
Twice a number	$2x$ or $x2$
Three-fourths of a number	$\frac{3}{4}x$ or $x\frac{3}{4}$
The product of 8 and a number	$8x$ or $x8$

TABLE 6.4

Division Key Words and Phrases

English	Mathematics
9 divided by a number	$9 \div x$ or $\frac{9}{x}$
A number divided by 9	$x \div 9$ or $\frac{x}{9}$
The quotient of 15 and a number	$15 \div x$ or $\frac{15}{x}$

TABLE 6.5

Mixed Translations

English	Mathematics
3 less than four times a number	$4x - 3$
6 more than two-thirds of a number	$6 + \frac{2}{3}x$
The square of the sum of a number and 5	$(x + 5)^2$

This means we can switch the order of the addition problem and still arrive at the correct answer. For example, $3 + 8 = 11$ and $8 + 3 = 11$.

The operation of subtraction, however, is not commutative. For example, $7 - 2 = 5$, but $2 - 7 = -5$. Therefore, subtraction must be carefully translated. In the second row, we are translating 12 less than "a number." What if the "number" was 30? Doesn't 12 less than 30 result in 18? More specifically, if there were 30 people in one room, and another room had 12 less people, wouldn't there be 18 people in the second room? The same applies to 21 subtracted from a number. Let us say the number was 35. Isn't $35 - 21 = 14$?

Conversely, $21-35$ does not equal 14. Additionally, this is why students in early elementary school may struggle more with subtraction than addition. Subtraction allows for less flexibility than addition, which can be problematic for students.

Like addition, multiplication is commutative. Doesn't $2(5)$ and $5(2)$ give us the same product (the answer to a multiplication problem), which is 10? Of course, it is common practice to write the number before the variable. In fact, that is why, for example, in the term, $6x$, we refer to 6 as the coefficient. As for the fractions, regarding "of," we discussed this in Chapter 5, but let us take a deeper look. What if there is a sale on a flat screen TV for half off? And that sale is for half off the list price of $300? Basically, you are paying half of the $300. How do you arrive at that? How about $\left(\frac{1}{2}\right)(300)$? Isn't that $150? Therefore, when taking a fraction "of" a number, we always multiply the fraction and the number.

Division, like subtraction, is not commutative. Additionally, "shared evenly" or "split evenly" also indicate division.

Of course, most applications require translation with multiple operations. See Table 6.5 for some examples.

To elaborate, on the third row, when a quantity is "squared" it is raised to a power of two. More specifically, the quantity is multiplied by itself.

Quantitative Personalities may Struggle

Students with quantitative personalities may struggle when there is not a precise formula for a specific problem. This can be the case with word problems. The reality is two different word problems can look very similar but have different ways of solving them. Consider the word problems in Figures 6.1 and 6.2.

Where do students get stuck? Again, students may not know where to begin. The student may panic and start writing down various formulas. Now, I am going to rewrite the two word problems, from Figures 6.1 and 6.2, but with basic directions as to how to solve them. More specifically, Figure 6.3 shows the word problem from Figure 6.1 re-written with specific directions as to how solve the problem. Figure 6.4 shows the word problem from Figure 6.2, which is also rewritten with specific directions as to how to solve it.

The application in Figure 6.1 requires the student to compute the perimeter of a rectangular pool:

Jessica is looking to put fencing around her pool. Her pool measures 20 feet by 22 feet. How much fencing would she need to place around her pool.

FIGURE 6.1

Jessica is looking for a pool cover. Her pool measures 20 feet by 22 feet. What size pool cover does Jessica need?

FIGURE 6.2

So that Jessica can obtain the correct fencing for her rectangular pool, find the perimeter if the length of the pool is 20 feet and the width of the pool is 22 feet.

FIGURE 6.3

So that Jessica can obtain the correct pool cover for her rectangular pool, find the area if the length of the pool is 20 feet and the width of the pool is 22 feet.

FIGURE 6.4

Perimeter of a rectangle $= 2(length) + 2(width)$. By the way, this translates to $2(20) + 2(22)$. This gives us a perimeter of 84 feet. However, to find the measurement of a pool cover, we need to find the area: Area of a rectangle $= length \times width$. This translates to $20(22)$. We arrive at an area of 440 square feet. However, students may confuse the formulas. In summation, word problems are much more challenging, as students do not receive specific directions as to how to start. This can be challenging for the quantitative thinker, solving a problem without a given formula or specific directions. It is noteworthy that these are only basic problems in arithmetic.

Lacking the Prerequisite Skills

Students' lacking prerequisite skills is a common link to math anxiety, and that applies to word problem applications as well.

> *The problem was even when I could figure out the word problem, I would get to the translation with the equation, and then I couldn't solve the problem.*
>
> Bob

> *It was so frustrating. It got to the point where I could follow my teacher step-by-step in doing a word problem. Heck, I even got to the point where I could set word problems up on my own, but when it came time to solving those problems, I just couldn't do it. I would look at the equations, and I would just get stuck.*
>
> Allison

A gardener is planting two types of flowers: tulips and daffodils, in a row. The number of tulips is 5 more than twice the number of daffodils. If there are 35 flowers in total, how many tulips and daffodils are there?

FIGURE 6.5
Introductory algebra word problem.

In this section, we will focus on how lacking proper prerequisite skills can hinder student success and create frustration. Let us review the application I listed at the start of this chapter, which is in Figure 6.5. It is an application from an introductory algebra class, and do not panic if you cannot solve the problem. If you have not completed basic algebra or an equivalent class, that is understandable.

Students can get stuck with translating this problem if they struggle with math terminology, and basically give up on the problem. So, let us take it step-by-step and use some of the translations from the tables earlier in the chapter.

- What are we solving for? We are solving for the number of tulips, and daffodils, so we need to assign some kind of value to each. Let us start with calling the number of daffodils, "x," because we know absolutely nothing about the quantity of daffodils. We know that the number of tulips is 5 more than twice the number of daffodils, so let $5 + 2x$ represent the number of tulips.

- Let us keep in mind that, in general, word problems translate into a for-mula (as we saw in Figures 6.1 and 6.2) or an equation. Looking at the second sentence, we can see there are a total of 35 flowers, so the number of daffodils plus the number of tulips is 35. So basically: $x + 2x + 5 = 35$. Our goal is to get x by itself, because that is how we identify the variable.

- Next, we must solve this linear equation. We start by combining the terms with x's, and get: $3x + 5 = 35$. Next, we subtract 5 from both sides:

$$3x + 5 = 35$$
$$-5 = -5$$
$$3x = 30.$$

Next, to get the x by itself, we divide both sides by 3, because the 3 is attached to the x by multiplication.

$$\frac{3}{3}x = \frac{30}{3}$$

In opening a new business, Alan, a new business owner, has calculated that the cost function is $C(x) = 70x + 2200$. The revenue function is $R(x) = 200x - x^2$. Determine how many items Alan must sell to break even. Additionally, at this time, Alan cannot produce more than 100 items.

FIGURE 6.6

Elementary/intermediate algebra word problem.

Therefore, $x = 10$. What were we solving for again? The number of daffodils and tulips and going back to the start we let x = the number of daffodils. Therefore, there are 10 daffodils. How about the number of tulips? We let the number of tulips be $2x + 5$, but we know that $x = 10$, so $2(10) + 5$, and that is 25. Therefore, the number of tulips is 25. So, there are 10 daffodils and 25 tulips.

Consider another word problem, in Figure 6.6, which is from either elementary or intermediate algebra or even a math for business analysis class.

The break-even value indicates the number of items a business must sell to be at exactly zero profit. More specifically, how many items must Alan sell so that the business is not losing any money but is also not making any profit. To determine the break-even value from cost and revenue equations, we simply set the equations equal to each other and solve for x, the unknown:

$$200x - x^2 = 70x + 2200.$$

The translation of this problem is not too difficult. In fact, this translation may be easier than the problem in Figure 6.5. However, once translated, students can get stuck in any number of places in solving the equation if they lack the prerequisite skills. First, the student must recognize that this is a quadratic equation. Why? Because the highest exponent in the equation is two, and if the student does not understand how to manipulate and solve a quadratic equation, this is where the frustration starts. So, where does the student go next? See below, and keep in mind that solving this problem involves several prerequisite algebra skills.

Step 1: Keep in mind that all quadratic equations have two solutions. Also, all quadratic equations must be set to zero, and we want the term to the second power to be positive.

$$\text{Therefore: } 200x - x^2 = 70x + 2200$$

$$-200x + x^2 = -200x + x^2.$$

Combining like terms, we get:

$$0 = x^2 - 130x + 2200$$

Step 2: We need to solve for x. For a quadratic equation, there are two ways we can do this. We can factor (as discussed in Chapter 5), or we can use the quadratic formula. We can use the quadratic formula to solve any quadratic equation. Since this is a rather complex quadratic equation, using the quadratic formula would be more efficient. The quadratic formula is as follows:

$$x = \frac{-b \pm \sqrt{b^2 - 4ac}}{2a}$$

This seems like a foreign language, right? Well, keep mind that "a," "b," and "c" represent numbers, and in a quadratic equation, "a" represents the number in front of the x^2 term, and that is 1; "b" represents the number in front of the x term, which is –130, and "c" represents the constant, which is 2200. So, one more time: $a = 1$, $b = -130$, $c = 2200$. Next, we will substitute the numbers for the variables and then simplify:

$$x = \frac{-(-130) \pm \sqrt{(-130)^2 - 4(1)(2200)}}{2(1)}$$

Step 3: Let us break this down to where we can see what "x" is.

$$x = \frac{130 \pm \sqrt{8100}}{2}.$$

More specifically, the double negative made the 130 positive; I used a calculator to simplify what is under the square root, and in the denominator, I multiplied the two values.

Let us continue to simplify. Keep in mind the square root of 8100 is 90, because 90(90) = 8100.

$$x = \frac{130 \pm 90}{2}.$$

This means $x = \frac{220}{2}$ and $\frac{130 - 90}{2}$.

These are our two answers:

$x = 110$ and $x = 20$.

So, those are our two answers? Wait, what were we solving for again? We were trying to determine how many items the business would need to sell to break even. Are there two answers to this question? Not exactly, because if you re-read the original problem, it states that Alan cannot produce more than 100 items. Therefore, we need to reject the solution of 110. In summation, Alan needs to sell 20 items to break even.

Bad Experiences Lead to a Self-Fulfilling Prophecy

In Chapter 1, I asked you to reflect upon poor prior experiences in math. I am sure many of those experiences had to do with word problem applications. Unfortunately, just the appearance of a word problem can send students into panic.

> *It just got to the point where I would read a word problem and freak out. I had no idea what to do, and the panic would just set in.*
>
> Alicia

> *I really tried to solve word problems, but I just went into them thinking about how much I sucked at word problems. I tried, but the negative thoughts just kept coming. It was like a little voice was saying to me, "You can't do this." I guess I just couldn't get out of my own way.*
>
> Peter

Sadly, this is case for many students. Additionally, students tend to panic and write anything down. Consequently, this leads to failure. This is an example of a self-fulfilling prophecy. Again, this is when expectations and beliefs toward a situation can influence behaviors and cause such beliefs to become true. Word problems in math can be difficult for everyone, even math teachers! However, you can successfully conquer word problems, and we will discuss how that is possible.

What are the Prerequisite Skills Needed?

As discussed earlier, the lack of prerequisite skills can be frustrating for students when attempting word problems. Moreover, the lack of prerequisites can be a barrier. Start your attack on word problems by addressing pre-requisite concepts. Before beginning any section involving word problem applications, ask your professor what concepts you need. These are the types of questions teachers love! More specifically, you may want to ask your teacher for examples of equations or expressions that you may need to know. You could do this in person or via email. Below is an example of a thoughtful and cordial communication.

Dear Professor Smith,
I know we will be starting word problems in class. I want to make sure
I am fully prepared. Could you let us know what equations, expressions,
or formulas we need to know to solve the word problems. Thank you so
much for your time.

For example, for the word problems in Figures 6.1 and 6.2, the student needs
to know the formulas for perimeter and area but also needs to understand
the concepts of perimeter and area. What are area and perimeter, and how
do they differ from each other? Prerequisite concepts for word problems may
be something some professors address anyway. In any event, as a student,
you need have an extremely specific understanding of the expressions and
equations needed to complete specific word problems.

Now, Attack the Problem!

Now, that we have addressed prerequisite skills and other areas students can
get stuck, let us set forth a full plan to attack math word problems with six
examples.

Example 1

Read the Problem

My guess is you are now rolling your eyes. As long as word problem
applications have been a part of mathematics education, teachers have been
instructing students to "read the problem." Additionally, as we discussed
earlier, students often draw a blank and panic when reading word problems.
However, let us take an innovative approach. Consider the following word
problem:

After reading the problem, understand that it is normal to lack clarity. You
may not be able to solve this problem immediately. The worst thing you can
do at this point is to begin to solve it for the sake of solving it while you still
lack clarity on how to proceed.

Ask Yourself some Questions

After you have read the problem, preferably twice, ask yourself the following
questions:

1. Do I understand what I am supposed to solve for or find?
2. Do I understand how to translate the information given, in the problem, to an expression or an equation to help me answer this?
3. What is still confusing me?

Regarding # 3, even math educators and mathematicians struggle with word problems. I can recall a terrific statistics professor, Joyce McQuade, that I had in college. One day she relayed to us that another professor had presented her with a problem from an advanced statistics class, and her response was, "I couldn't do it." I was shocked. Professor McQuade was this brilliant mathematician and stellar teacher, and she could not solve a problem? I can understand that now. As a math educator, I constantly try new applications that I find online to keep myself sharp. There are times I will run into problems that give me difficulty. I either need to conduct additional research or ask a colleague for help. The reality is word problems have a difficulty and abstractness that computational problems do not. Even experienced math educators need help. The key is, like anything else in math, to focus in on your question. Reflect on the following questions: Do you understand what you need to solve for? Is it that you understand what you need to solve for but are having difficulty translating the problem? Is it that you can translate some of the problem but not all of it? Narrowing down your questions will not only help your teacher help you, but this process will help you feel better. Rather than simply feeling as if you "do not understand anything," which can send students into a panic, you will feel more in control of the situation.

Understand the Prerequisites

For this word problem, we need to understand:

- How to translate words to multiplication and addition.
- How to solve a linear equation.

Identify What You are Solving For and Translate

Referring to the problem in Figure 6.7, we are solving for the number of hours that Mike spent babysitting. That is some number that we do not know.

Mike earns $6 an hour babysitting for his younger sister. His mom gave him $71 last week. This included his babysitting money and his $20 allowance. How many hours did Mike babysit?

FIGURE 6.7
Example 1 for word problems.

1. We will let x = the number of hours that Mike spent babysitting. More specifically, we are looking for one answer.
2. Next, we need to translate the information given into an equation to help us solve. If you are looking for one answer, you need only one equation.
    a) In the equation that you need, do you know the total amount of something? You know that Mike received a total of 71 dollars. You could start by writing: some equation or something with an "x" = 71. We have no idea what will equal 71, but we have a start.
    b) Can you translate the rest using the math terminology we discussed earlier? Mike is earning $6 an hour, so we need to multiply 6 by the hours. Since we allowed x to equal the number of hours that Mike spent babysitting, this would translate to $6x$. Is that all? Mike also received $20 for his allowance. That is, $20 added on to the money he earned babysitting. Therefore, the equation translates to: $6x + 20 = 71$. Let us recap. Mike earned $6 an hour babysitting. That means six times the number of hours babysitting is the total amount of money earned. Since, we do not know how many hours he spent babysitting, it is x. However, he also received a fixed amount of $20 on top of babysitting money, which is why we add 20. Again, $71 is the total amount earned; therefore, the 71 is after the equal sign. Therefore, the problem translates to: $6x + 20 = 71$.

Solve the Problem

This involves isolating for and solving for x:

Step 1: Subtract 20 from both sides: $6x + 20 = 71$
$$-20 = -20 \text{ to get: } 6x = 51$$

Step 2: To isolate, divide both sides of the equation by 6: $\dfrac{6x}{6} = \dfrac{51}{6}$

Our solutions is 8.5, and more specifically Mike spent 8.5 hours babysitting.

Does this Answer Make Sense?

It is common for students to arrive at an answer involving fractions or decimals and assume they are incorrect. However, does it make sense that this answer could be a decimal? The answer is in terms of hours. Can hours be measured in decimals? Yes. Furthermore, does the value, 8.5 make sense? Keep in mind, Mike is making $6 per hour and only made a total of $71. Therefore, his total number of hours spent babysitting would not amount to much. If we had arrived at an answer such as 40 or 50, we should instinctively know that something is wrong.

Example 2

See Figure 6.8 for a problem from either elementary algebra, intermediate algebra, or possibly even introduction to statistics.

> In his math class, Dan received exam grades of 55, 75, 80, and 82. His final quarterly grade will be the average of five exams. What grade will Dan need to receive on his fifth exam so that he receives at least a "C" in the class? Please note that a final average of 70% is the lowest grade for a "C."

FIGURE 6.8
Example 2 for word problems.

Read the Problem

Ask Yourself Some Questions

1) Do I understand what I am supposed to solve for or find?
2) Do I understand how to translate the information given, in the problem, to an expression or an equation to help me answer this?
3) What is still confusing me?

Understanding the Prerequisites

For this word problem, we need to understand:

- How average (or mean) is computed.
- How to translate an application into a fractional (rational) equation.
- Solving a proportional equation.
- Solving a linear equation.

Identify What You are Solving For and Translate

1) We will let x = the grade Dan must receive.

2) Since we need one answer, we need one equation.

3) We know that this problem involves using the average, which is when we add the values and divide by the total number, and that number would be 5, as there are five values. We also know Dan wants to achieve a 70; therefore, the 70 should be on the other side of the equal sign. So, we need to add the five exam grades and divide by 5, and we should obtain 70, but again, the fifth exam grade is "x," as we do not know it and need to solve for it. Therefore, the problem translates to the following equation:

$$\frac{55+75+80+82+x}{5} = 70$$

Solve the Problem

We need to isolate and solve for x.

Step 1: Let us simplify the numerator of the fraction by adding what we can, and we get:

$$\frac{292+x}{5} = 70$$

Step 2: It is very difficult to solve for a variable if there is a fraction. Therefore, the best way to solve is to remove the fraction. Since this is a proportion, which is two fractions that are equal (don't forget that we can express 70 as $\frac{70}{1}$), we can solve a proportion by cross multiplication. We can cross multiply $(292+x)(1)$ and $(5)(70)$. That gives us: $292+x=350$.

Step 3: Let us solve this linear equation: by subtracting 292 from both sides to isolate the x

$$292+x = 350$$

$$-292 \quad = -292$$

$x = 58$, which means that George must score at least a 58 to receive a grade of "C."

Does this Answer Make Sense?

First, you can check your answer. Do the grades of 55, 75, 80, 82, and 58 average to 70? Add those five values and divide by 5, and we can see this is correct. Additionally, considering these numbers must average to 70, we can see a balance of grades in the 50s and 80s that would likely average to a 70.

Example 3

For Examples 3 and 4, let us apply the same process with different types of word problems focusing on probability, a topic from either quantitative reasoning or introduction to statistics. We discussed probability as a troublesome topic for students in Chapter 5. These two examples give us more of an opportunity to dive into this topic.

Read the Problem

Two fair dice are rolled. What is the probability of obtaining a six on the first die and an even number on the second die?

FIGURE 6.9
Example 3 for word problems.

Ask yourself the questions from Examples 1 and 2.

Understand the Prerequisites

For this problem, we need to understand:

- What is a "fair die"? A fair die is a die where every side is of equal size. If a fair die has six sides, which most do, we have the same chance of the die landing on a given side when rolling.
- How many events are there in this situation?
- The formula for "and" in probability.

Before Translating, Draw a Picture or Diagram (Sometimes)

As discussed earlier, quantitative thinkers tend to like to rely on algorithms and procedural processes, and this certainly applies to word problems. However, qualitative thinkers like to visualize, and quite frankly, visualizing can be helpful in probability. Therefore, before translating, a diagram may be helpful. For probability applications, this is known as the sample space where we list all the possible outcomes. We know that the ultimate

1, 1	2, 1	3, 1	4, 1	5, 1	6, 1
1, 2	2, 2	3, 2	4, 2	5, 2	6, 2x
1, 3	2, 3	3, 3	4, 3	5, 3	6, 3
1, 4	2, 4	3, 4	4, 4	5, 4	6, 4x
1, 5	2, 5	3, 5	4, 5	5, 5	6, 5
1, 6	2, 6	3, 6	4, 6	5, 6	6, 6x

FIGURE 6.10
Possible outcomes for rolling two dice.

answer to a probability question is the $\dfrac{number\ of\ desired\ outcomes}{total\ number\ of\ outcomes}$. Figure 6.10

lists all the possible outcomes for the word problem in Figure 6.9. I placed an x next to each outcome that we want (a six on the first roll and an even number on the second roll). More specifically, "1, 1" means we can roll a 1 on the first die and a 1 on the second die; "1, 2" means that we could roll a 1 on the first die and a 2 on the second die, and so forth. We still need to identify what we are looking for, which is the probability that we achieve six (on the first die) and an even number (on the second die)., which is $\dfrac{number\ of\ times\ we\ want\ to\ achieve\ a\ six\ and\ an\ even\ number}{total\ number\ of\ outcomes}$. Keep in mind that "and" means that both events must happen, rolling a six and rolling an even number.

We can see that there are 36 possible outcomes. If we add all the x's, which are beside the desired outcomes, we obtain 3, or in probability form, $\dfrac{3}{36}$. More specifically, $\dfrac{number\ of\ times\ we\ want\ to\ achieve\ a\ six\ and\ an\ even\ number}{total\ number\ of\ outcomes} = \dfrac{3}{36}$.

Identify and Translate

When solving word problems, using visuals or diagrams is important for any math student, as such visuals can better contextualize a problem and provide a deeper understanding. However, oftentimes visuals become too complex. For example, what if the probability problem in Figure 6.9 contained over 100 outcomes? Would it be reasonable to write out over 100 outcomes? Of course not. That is why we rely on algorithms.

1.   Again, our x = the probability of obtaining a 6 and an even number.

2. For a probability problem, we should always identify our events. In this case there are two events: obtaining a 6 and an even number. Let us label these two events and list the probabilities:

   $P(A) = \dfrac{1}{6}$, obtaining a 6. More specifically, this is the probability of the "A" event. This is because the probability of obtaining a 6 is $\dfrac{1}{6}$.

   $P(B) = \dfrac{3}{6}$, obtaining an even number. More specifically, this is the probability of obtaining the "B" event. This is because the probability of obtaining an even number is $\dfrac{3}{6}$.

   In reading the problem, we see the desired outcomes connected by "and," which indicates that we multiply the probabilities of the two events. $P(A) \cdot P(B)$.

Solve the Problem

$$\frac{1}{6} \cdot \frac{3}{6} = x$$

The probability that we obtain a 6 on the first die and an even number on the second die is $\dfrac{3}{36}$, and in reduced form this is $\dfrac{1}{12}$.

Does the Answer Make Sense?

Considering that probability is always one or less, or 100% or less, this answer is within our domain (allowable answers).

Mutual Exclusivity

To best understand probability, students must understand the concept of mutual exclusivity. More specifically, the student must understand the terms, "mutually exclusive" and "not mutually exclusive." Think of two events that cannot happen at the same time:

If you roll a fair die just one time, what is the probability it lands on a 3 and a 5?

No, that is impossible. The die can only land on one number. Therefore, those two events are mutually exclusive. Therefore, the answer to the question above is "no solution" or "not possible." However, the probability of obtaining a 6 on one die and an even number on the other die can happen at the same time. Therefore, those events are not mutually exclusive. My point is that we cannot assume that simply because two events are connected by "and" that we can multiply the events. We can only multiply if the events are not mutually exclusive.

Example 4

See Figure 6.11 for another example focusing on probability.

Understand the Prerequisites

For this problem, we need to understand:

- What is a "fair die"?
- How many events are there in this situation?
- The difference between mutually exclusive (events that cannot happen at the same time) and not mutually exclusive (events that can happen at the same time).
- The formula for "or" in probability, otherwise known as the General Addition Rule.

Before Translating, Draw a Picture or Diagram (Sometimes)

Like Example 3, let us draw a diagram of all the possible outcomes (see Figure 6.12). For example, you could roll a 1 and then heads, a 2 and then heads, a 1 and then tails, and then so forth. Again, there is an x listed next to

You roll a fair die and then toss a coin. What is the probability of obtaining an even number or heads?

FIGURE 6.11
Example 4 for word problems.

1, H x	1, T
x 2, H x	x 2, T
3, H x	3, T
x 4, H x	x 4, T
5, H x	5, T
x 6, H x	x 6, T

FIGURE 6.12
Possible outcomes for rolling a die and tossing a coin.

the desired outcomes which are obtaining an even number or heads. Keep in mind that "or" means that we are happy with either an even number or heads. We can see that there are 12 outcomes. If we add all the x's, which are beside the desired outcomes, we obtain 12, or in probability form, $\frac{12}{12}$. However, there are three outcomes that contain two x's or a double desired outcome. Therefore, we must subtract three, and we arrive at $\frac{9}{12}$. Think about it. If you win a prize for landing on an even number or heads, you can only win a prize nine possible times. If the die lands on 2 and the coin lands on heads, you will not win two prizes! You will still only win one prize.

Identify and Translate

Again, diagrams help us understand the concepts, but it is better to use algorithms, as problems get more complex. That is why we rely on algorithms.

1. Again, our x = the probability of obtaining either an even number or heads.
2. For a probability problem, we should always identify our events. In this case there are two events: obtaining an even number and obtaining heads. Let us label these two events and list the probabilities:

    $P(A) = \frac{3}{6}$, obtaining an even number. More specifically, this is the probability of the "A" event.

    $P(B) = \frac{1}{2}$, obtaining heads. More specifically, this is the probability of obtaining the "B" event.

    In reading the problem, we see the desired outcomes connected by "or," which indicates the General Addition Rule of probability. However, we know that there are two General Addition Rules:
    * $P(A) + P(B) = x$ (use when events are mutually exclusive)
    * $P(A) + P(B) - P(A \cdot B) = x$ (use when events are not mutually exclusive)

Which rule do we select? Again, this is where we need to understand the prerequisite concept of mutual exclusivity. Can events A and B happen at the same time? Since it is possible to obtain an even number and heads at the same time, the events are not mutually exclusive; therefore, we use the second formula.

Solve the Problem

$$\frac{3}{6}+\frac{1}{2}-\left(\frac{3}{6}\cdot\frac{1}{2}\right)=x$$

The probability that we obtain either an even number on the die or heads from the coin is $\frac{3}{4}$ or if we convert to a percent (by performing $3 \div 4$ and multiplying by 100), 75%.

Does the Answer Make Sense?

Considering that probability is always one or less, or 100% or less, this answer is within our domain (allowable answers).

Example 5

See Figure 6.13 for another problem for elementary or intermediate algebra.

Read the Problem

A store sells two types of candies: chocolate bars and gummy bears. The chocolate bars cost $2 each, and the gummy bears cost $1 each. A customer buys 30 pieces of candy for $50. How many chocolate bars and gummy bears did the customer buy?

FIGURE 6.13
Example 5 for word problems.

Ask Yourself Some Questions

Ask yourself the questions listed in the previous examples. We will list the prerequisites below.

Understand the Prerequisites

For this problem, we need to understand:

- How to translate a problem into two variables.
- Solving a system of two equations.

Identify and Translate

1) There are two quantities we need to solve for in this problem, the number of chocolate bars and the number of gummy bears. However, it is not possible to solve this problem in terms of one variable, like we did for the problem in Figure 6.5.

2) This problem requires us to use two separate variables for the two quantities. Therefore, we will allow x = the number of chocolate bars, and y = the number of gummy bears.

3) Whenever we use two variables, we need to use two equations. In fact, the number of variables always dictates the number of equations. What will our equations be? In reading the problem, there are two situations: the number of chocolate bars and gummy bears, and the cost of chocolate bars and gummy bears. Consequently, the first equation will focus on the number of chocolate bars and gummy bears, and the second equation will focus on the cost of each.

4) Here is the translation:
$x + y = 30$ (this is because we know there are 30 total candies)

$2x + y = 50$ (this is because each bar is $2; each bear is $1, and the total is $50)

Solve the Problem

We can solve a system of equations by either elimination or substitution. Since the x's and y's are lined up, we will solve by elimination. This means we will eliminate one variable and solve for the other. Let us start by eliminating the y. We need to ensure that the y's cancel out. Since the y has a coefficient of 1, we can multiply the top equation by 1 and the bottom equation by –1.

(1) $x + y = 30$

(–1) $2x + y = 50$. This gives us:

$x + y = 30$

$-2x - y = -50$. Now, we will add vertically, combining like terms.

$-x = -20$. We solve for x by dividing both sides by –1.

$$\frac{-1}{-1}x = \frac{-20}{-1}$$

Therefore, $x = 20$. But what is 20? Since we let x = the number of chocolate bars, there are 20 chocolate bars. What about y? Since we know what x is, we can substitute 20 for x into either equation.

$20 + y = 30$. Now we solve for y by subtracting 20 from both sides:

$20 + y = 30$

$-20 \quad = -20$

Therefore, $y = 10$. So, there are 20 chocolate bars and 10 gummy bears.

Does the Answer Make Sense?

We can check our answer. We should be able to substitute 10 for x and 20 for y into either equation and arrive at either solution. Let us do that:

$20 + 10 = 30$

$2(20) + 1(10) = 50$

Both equations check!

Example 6

For this last example (see Figure 6.14), we are going to look at an application from trigonometry. This type of problem is for students who will be on the STEM (Science, Technology, Engineering, and Mathematics) path, and more importantly will be on the path to take calculus. You may never need a problem like this, or if you are on the STEM path, you can always come back to it. However, I wanted to demonstrate how this problem-solving process can work for more complex applications as well.

Read the Problem

From a point on the ground 45 feet from the foot of a tree, the angle of elevation of the top of the tree is 35°. Find the height of the tree.

FIGURE 6.14

Example 6 for word problems.

Ask Yourself Some Questions

Ask yourself the questions listed in the previous examples, and if you are still confused ensure that you have the proper prerequisite skills to complete this problem.

Understand the Prerequisites

For this problem, we need to understand:

- The concept of translating words into a right triangle.
- The rules of right triangle trigonometry: finding the sine, cosine, and tangent of an angle.
- The idea of the angle of elevation from a given angle.
- Solving proportions.

Identify, Translate, and Draw a Picture

In the previous examples, I mentioned that it can be helpful to write out a visual or diagram. However, there are some cases where a visual is mandatory, and that is the case for applications in more advanced geometry and trigonometry. When drawing a picture is imperative, we need to incorporate that into our identification and translation.

1. We can infer that this is a right triangle. So, start by drawing a right triangle (see Figure 6.15).
2. We know that x = the height of the tree because that is what we are solving for. We also know that the foot of the tree forms the right angle with the ground. Consequently, the distance from the foot of the tree to the third point in the triangle is 45 feet. Additionally, the angle of elevation from the top of the tree is 35°. More specifically, we need to look downward at an angle from the top of the tree. All this information leads to labeling the right triangle in Figure 6.16. And please do not laugh at my tree!
3. Now, we need to translate this to an equation or a proportion. Typically, right-angle trigonometry translates to proportions. We need to use either the sine, cosine, or tangent function to try and solve. Keep in mind that in a proportion, we can only have one unknown. The sine of Angle A, which is the opposite side over the hypotenuse would not help, because we do not know either the opposite side or the hypotenuse. That would translate to: $\sin(35) = \dfrac{x}{blank}$. There are two unknowns there. The cosine of Angle A, which is the adjacent side

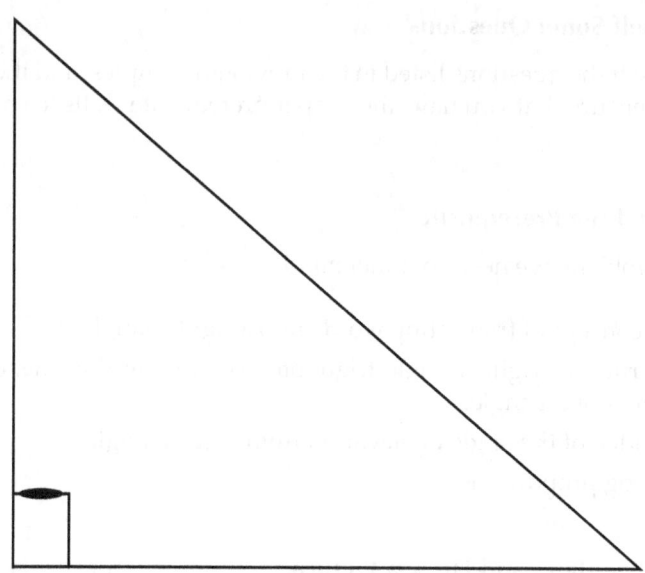

FIGURE 6.15
Right angle

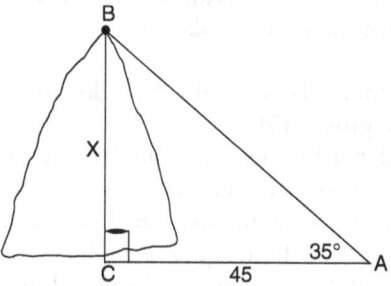

FIGURE 6.16
Right angle with information

over the hypotenuse would translate to $\cos(35) = \dfrac{45}{blank}$. There is only one unknown there, but nothing there involves the side of "x," which is what we are solving for. However, the tangent of Angle A is equal to the opposite side over the adjacent side, which is, $\text{Tan}(35) = \dfrac{x}{45}$.

Solve the Problem

We can solve any proportion by cross multiplication. Using at least a scientific calculator, we can multiply $\text{Tan}(35)(45)$ and set that equal to x multiplied by 1. Keep in mind that the $\text{Tan}(35)$ has a 1 underneath it.

$$\text{Tan}(35)(45) = x$$
$$31.5 \text{ feet} = x$$

Does the Answer Make Sense?

31.5 feet makes sense as the height for a tree.

Struggling?

As I mentioned earlier in this chapter, everyone (students and teachers) struggles with word problems at times. While being prepared and having a plan helps, there will be times you will still struggle with a word problem. What can you do?

Look Over your Class Notes

There is a good chance your professor has covered sample problems in class that resemble the current word problem. In reviewing the guided practice in class, you may see something in the setup or the translation that may help.

Is There an Online Tutorial?

As mentioned earlier, college math classes often utilize math software programs such as My Math Lab, ALEKS, or Hawkes. Such programs embed video tutorials in each section. Try watching a video to determine if it sheds light on your question.

Ask your Instructor... Right Away!

Do not wait. Get help with this problem immediately. Visit your professor during an office hour. Again, be sure to pinpoint exactly where your question is.

It Works!

I never realized I could do word problems! I got through my pre-algebra class and my [introduction to] statistics class. It was a matter of just calming down and having a plan and taking it step-by-step. It was also a matter of getting help when I need it.

Alicia

Once I started getting word problems right in my algebra class, it started clicking. I just got more confidence that I could do word problems. Could I always get them right? No. Sometimes I needed help, but knowing what I needed to know before the problem [the prerequisites] really helped.

Allison

Why in the World? Part 6

Why in the World do we Flip the Sign in an Inequality When Dividing by a Negative Coefficient?

Please see Table 6.6 where we compare solving two inequalities.

Why this makes no sense: The inequality on the left side makes sense; we solved it just as we solve linear equations, but the inequality on the right side makes no sense whatsoever. Why did we flip the direction of the sign just because we divided by a negative coefficient? That looks like a magic trick!

TABLE 6.6

Inequalities

Solve: $2x - 5 > 11$, add 5 to each side	Solve: $-2x - 5 > 11$, add 5 to each side
$+5+5$	$+5+5$
$2x > 16$, divide each side by 2	$-2x > 16$, now divide by -2
$\dfrac{2}{2}x > \dfrac{16}{2}$	$\dfrac{-2}{-2}x > \dfrac{16}{-2}$
The answer is $x > 8$	The answer is $x < -8$
This means x can be any number greater than 8.	This means x can be any number less than -8.
We can also write this as $(8, \infty)$. This means x can be any number greater than 8 but less than infinity.	We can also write this as $(-\infty, -8)$. This means x can be any number greater than negative infinity but less than -8.

Let us try to understand: To make some sense of this bizarre rule, let us go back to the basics of inequalities, which we have worked with since elementary school. Remember this: $10 > 6$? You may not have realized it at the time, but this is an inequality. This consists of two quantities that are not equal. How about this? Just for fun, let us divide the 10 and the 6 by 2: $\frac{10}{2} > \frac{6}{2}$ and we get: $5 > 3$. We arrive at two different quantities, but it is still a true inequality. Now, let us divide both sides by a -2 and not change the direction of the inequality (as we are supposed to). $\frac{10}{-2} > \frac{6}{-2}$. We get: $-5 > -3$. Is this true? Is -5 greater than -3? If you owe $5, are you in better financial shape than if you owe $3? No. Ultimately, when we divide an inequality by a negative, we change the relationship. Let us take a closer look at a basic number line (see Figure 6.17).

On the right side (right from 0) of the number line, the farther the numbers are away from 0, the larger they are. On the left side (left from 0), the farther the numbers are away from 0, the smaller they are. By dividing both by a negative, each quantity moves to the other side of 0, which changes the relationship.

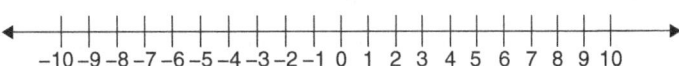

FIGURE 6.17
The number line

Exercises to Help You Practice

Below are some word problem applications to help you practice and employ the strategies we discussed in this chapter. Feel free to try some before you start your college classes or as extra practice while you are completing your college classes. The answers are in Appendix D.

Basic Algebra Word Problems

1) A bakery made 120 cookies to package in boxes. If each box holds eight cookies, how many boxes can they fill?

2) Emma is saving money to buy a new bike. She already has $20 saved, and she plans to save an additional $5 each week. How many weeks will it take for her to save a total of $70?

3) George has baked a total of 78 of two types of cookies: chocolate chip and oatmeal raisin. The number of chocolate chip cookies is two less than three times the number of oatmeal raisin cookies. Find the number of each cookie.

4) A car rental company charges a $50 base fee plus $10 per hour for a rental. If Sarah paid $80 in total, for how many hours did she rent the car?

5) The perimeter around a rectangular garden is 190 feet. The width is 5 feet more than twice the length. Find the length and the width.

6) Five subtracted from two-thirds of a number is 7. Find the number.

7) Marsha's report card grade in her math class will consist of the average of her seven quiz grades. Her first six grades are 87, 89, 90, 95, 94, 82. Marsha wants to achieve an "A" in her math class, and the lowest average to achieve an "A" is a 90. What grade does Marsha need on her seventh exam to achieve at least an "A?"

More Advanced Algebraic Word Problems

1) A ball is thrown into the air from a height of 5 meters with an initial upward velocity of 20 meters per second. The height (h) of the ball (in meters) at any time (t) (in seconds) after it is thrown can be described by the equation: $h(t) = -5t^2 + 20t + 5$. After how many seconds will the ball hit the ground?

2) The length of a rectangular yard is 216 square feet. The length is three feet more than three times the width. Find the length and the width.

3) The product of two numbers is 260. The larger number is 4 less than three times the smaller number. Find the two numbers.

Some Basic Probability

1) Consider the spinner below:

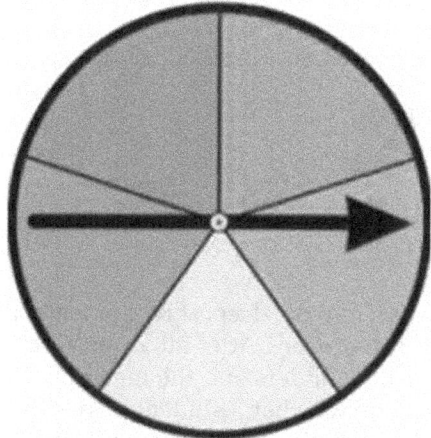

What is the probability that the spinner lands on the light shade?

2) A fair die is rolled once. What is the probability of obtaining a 3 or a 6?

3) A fair die is rolled and a coin is tossed. What is the probability of obtaining an odd number on the die and a tails on the coin?

4) A fair die is rolled and a coin is tossed. What is the probability of obtaining a 5 on the die and heads on the coin?

Activity

- Reflect on word problem applications that have given you difficulty in the past.

- Can you recall the reasons you struggled with these word problems? Did you lack the prerequisite skills? Were you unable to determine what you were supposed to solve for? Did you struggle in translating the problem to an equation?

- Earlier in this chapter, I discussed how students often panic when facing word problems. What can you do to change your mindset toward word problems? Do the strategies discussed in this chapter give you more confidence?

7

Tackling Exams

If we were to take a dive into the deepest root of math anxiety, we would see the fear of exams or test anxiety. After all, exams are where students must demonstrate whether they understand the material, and instructors use the exam scores to determine whether a student passes the math course. In this chapter, we are going to examine test anxiety and how to combat test anxiety.

Test Anxiety? You are in Good Company

At this point, you should realize that your math anxiety as well struggles within math classes are not uncommon. The same goes for test-taking anxiety. In fact, research has shown that 40–60% of students, ranging from kindergarten through college, are negatively impacted by test anxiety (Adelman & Taylor, 2010). However, those are just numbers. Let us hear some testimonials regarding test anxiety in math:

> *It got to the point in high school, where on the morning before exams I couldn't stop throwing up. Just the thought of being in the classroom and looking at the exam made me sick. Then, it would get to the point of where I was so out of control, I needed to talk myself out of throwing up.*

> Bob

> *I knew exactly what I was afraid of. It was looking at the test and freezing and sitting there while I could hear all those busy pencils and pens around me, other students just finishing the test, while I sat there in a panic, not knowing what to do.*

> Derek

> *Taking a test is what kept me from going back to school. Sitting in a classroom, especially as an old guy, just not knowing what to do was like something out of my worst nightmare.*

> Otis

DOI: 10.1201/9781003614142-7

I'm a person who likes to feel a sense of control in my life, and with exams, espe-cially math exams, It felt like everything was spinning out of control. My mind would just go blank.

Daniele

Can students feel test anxiety when taking an online test in the comfort of their own homes?

I thought taking an online class would make my test anxiety go away, but it was still there. I could feel the web cam right on my face with the time dwindling.

Alicia

The first test I took online, I couldn't download Respondus, so I couldn't take the test. I had an all-out panic attack.

Linda

Respondus is a tool utilized by many institutions for online exams. Respondus Lockdown Browser prohibits students from accessing any other website, application, or function on their computer screen while taking an exam. Respondus Monitoring is a proctoring tool that uses a webcam to flag any suspicious behavior while taking the exam. Students must download Respondus onto their computer, laptop, or iPad.

Root Causes of Test Anxiety

What are the causes of test anxiety? How do exams become such a harrowing experience for students? The reasons vary, but there are some overall causes.

Poor Test History

Like bad experiences in math, horrid test experiences haunt students.

I guess it kind of snowballed over the years. I had so many bad experiences taking tests that I just expected to freeze up and fail.

Joanne

It got to the point where even my parents would remind me of how I just was a bad test-taker. It just became who I was.

Brenda

Fear of Failure

The fear of failure can be daunting for some students. This can especially be the case for Generation Z (students born between 1995 and 2010) students (Seemiller & Grace, 2016). Failure is difficult for most people, but research has shown that Generation Z has an especially intense aversion to negative events, and that includes failure.

> *It was just that fear of feeling like a failure. I mean, I get a test back with a failing grade, and that's how I see myself. I failed the test, so I am a failure. It makes me cry, and I could just see it happening.*
>
> Audrey

> *So many times, I would get a test back with a failing grade, and I just felt like a failure. I failed the test, so I was a failure. It's the worst feeling in the world, and I just became so filled with fear out of feeling like that.*
>
> Emily

> *I've had teachers who used to treat me like the grades I would get on my math tests. Since my grades were garbage, they treated me like garbage. They were nice and gave all the attention to the kids who got A's.*
>
> Peter

High-Stakes Exams

For many students, the higher leverage the exam, the more anxiety the exam produces. More specifically, when an exam monopolizes a large portion of the course grade, or the student's final grade is dependent on passing the exam, the student feels a great deal of anxiety. This is in contrast to lower-stakes assessments such as quizzes or graded homework assignments.

> *It's so hard when you know that so much of your grade rides on the exam. It's like I keep thinking if I don't pass this exam, I'm going fail the class, and my life's over.*
>
> Daniele

> *In this one math class, we had to pass the final in order to pass the class. I mean, who comes up with that? I fail the final, and I fail the class. That made me freak out on the final.*
>
> Alicia

> *When it's a little quiz that doesn't count that much, it doesn't scare me as much, but the more that I know the test counts for my final grade, the more nervous I get.*
>
> Bob

High-stakes exams are part of life. In fact, most professions require students to pass an exam to enter that profession. More specifically, to enter professions such as accounting, real-estate, nursing, teaching, or any of the medical professions, a student must pass some sort of comprehensive exam to obtain licensure. Consequently, students must be able to pass high-stakes exams.

Poor Test Preparation

Finally, students experience test anxiety because they know they are unprepared or simply feel unprepared.

> *I get to the test, and I just feel like I'm not ready. I've studied, but I start questioning if I studied enough. I start doubting if I'm ready.*
>
> David

> *I'll admit it. There are so many times I wasn't ready for a test, and I would just get that panicked feeling because I knew I wasn't ready and was going to fail. I guess I brought it on myself, but it's why I freak before exams.*
>
> Derek

> *I think my biggest problem was I had no idea how to study for a math test. I knew how to study for a history test, a psychology test, but how do you study for a math test? Because I didn't know how to study, I didn't feel prepared, and that made me panic.*
>
> Peter

Tackling an Exam Starts Long Before the Exam

Students may attempt to study for a math exam; however, they wait too long to prepare.

> *My first two times trying to pass a math class [in college], I would miss class, not take notes, not do my homework, and then try to cram for the exam a couple of days before. I would try to do the practice test and just freak out, because I didn't know anything.*
>
> Tom

> *The people in the tutoring center really started to hate me. That's because I would show up there the day before the test with the study guide, and I would expect the tutors to just teach me everything.*
>
> Nancy

> *I would meet with my tutor right before an exam, and he would just roll his eyes at everything I didn't know. He would ask, "Why didn't you get help on this topic when you covered it in class." Honestly, I don't know.*
>
> Jessica

Math is so much different than other subjects. In a history class or a psychology class, you can catch up on material that you missed by reviewing before an exam. But in a math class, you have to really understand the material as you learn it. That's the difference, in math it's not about reading or memorizing, you have to truly understand what you're doing.

Donna

The bottom line is proper preparation for a math exam starts on the first day of class. We will discuss ways to prepare for an exam when it approaches as well as test-taking strategies; however, below are some tips to help you prepare and avoid shock or panic when it comes close to test time.

- Most professors post an outline of what topics are covered on specific days as well as due dates, and this includes exams. Make a note of the date of your exams, just so they do not sneak up on you.

- Whether it is homework or practice problems, ensure that you work on your math every day or almost every day. As we discussed in Chapter 4, the longer the lapse in time, the higher percentage of your math you will forget.

- At the end of each week, go through your homework or class problems for that week, and make note of any that you still struggle with or do not understand. Ensure that you get help with these problems immediately.

- What sections will be covered on the exam? For example, is it all of Chapter 3, which includes Chapter 3 sections 1–8? As you complete each section, make some notations as to which problems or topics give you difficulty.

- Take time to learn about your first exam. How much time will you have? More specifically, will you have the entire class or only part of the class? Keep in mind, if you receive special accommodations, you may have extended time. How many questions will be on the exam? Keep in mind that some professors do not wish to share this ahead of time, and you need to respect that.

Positive Thoughts

As you move toward your first exam, here are some statements to continue to tell yourself and even write down.

- My past exams do not define who I am. They have no influence on me. They do not control me.

- I am prepared for this exam, and I am in control.
- I can excel on this exam.
- I am an intelligent person.
- I am going to take this exam and see the types of questions I have been practicing, and I will successfully complete them.

How do I Study for a Math Exam?

You have arrived at the time (three to five days) before the exam. How do you study for a math test? As the students mentioned above, being successful in math is not simply about memorization but about understanding the content.

Is There a Study Guide?

Some professors will provide a study guide or practice test before the actual exams. These instruments provide students with a summary of the required material for the exam. Be sure to complete these study guides at least twice. Make note of any problems in which you have difficulty and get help immediately.

What if your professor does not provide a study guide? After all, providing a study guide is the professor's choice. Design your own study guide. Review the problems in each section and create a study guide based on the problems you believe will be on the exam. You can select these problems from your textbook and the online software.

Clean up any Mistakes

As mentioned earlier, you should keep track of topics that give you difficulty each week and in each section. Make it a point to get help with these problems and practice the types of problems that have given you difficulty. Additionally, make a list of the errors you have made and how you will go about resolving this. See Table 7.1 for three different examples. The first example is from basic algebra, the second is from intermediate algebra or college algebra. The third is from basic statistics.

By tackling the problems that have given you trouble, you will not only be able to solve these problems on the exam, but your confidence level will increase.

TABLE 7.1

Confronting Errors

Question	My Wrong Answer	What I did Wrong and What I Need to Fix
Simplify: $(6x^3y^4)^2$	$36x^9y^{16}$	I made a mistake with the exponents. I raised the 3 and 4 to the second power. I should have used the power rule, which is to multiply the exponents. I should have gotten: $36x^6y^8$
Find the domain for the following: $y = \sqrt{x-7}$	I set $x - 7 = 0$. I solved for x and got $x = 7$	Domain means all numbers that x is allowed to be for a given equation. Since the values under radicals can only be zero or higher, x can only be some number that gives us zero or higher. I should have set: $x - 7 \geq 0$. I would add 7 to both sides and get: $x \geq 7$. So, x can be any number from 7 and on. I should write this as $[7, \infty)$. This means x can be any number greater than or equal to 7 or less than infinity
Find the median for the following numbers: 5, 6, 10, 12, 15, 18, 20, 23	15	I got confused. I didn't realize that when I have an even amount of numbers (there are 8 numbers here), to find the median, I take the average of the two middle numbers. The two middle numbers are 12 and 15, so the median is 13.5.

Data Drop Off (Memory Dump)

The most frustrating thing for me when taking a test is how I would forget every-thing. I would look at the test and just panic. It's like I couldn't even remember my own name.

David

I swear there were times I was prepared for the test, but I just froze when I got the test. I looked at the first question and just went into a total panic. I couldn't remember anything. I couldn't understand why that was happening to me.

Louis

I would look at a test and my heart would just beat louder than I could think, and I would feel myself just sweating everywhere.

Emily

Has this happened to you when attempting an exam? You feel panic to the point of forgetting everything? Have you even exhibited the physical symptoms that Emily described? This is normal. Your body is experiencing something that psychologists refer to as the "fight or flight response." As human beings, when faced with a dangerous situation, our bodies are preparing to either fight for our lives or run from the situation. When facing a dangerous situation, our bodies are reacting to acute stress. Our bodies react by producing hormones such as the adrenocorticotropic hormone and corticotropin-releasing hormone. Consequently, this increases symptoms such as increased heartrate, sweating, flushed skin, and memory loss (Goldstein, 2010). The bottom line is when you receive your test, your body reacts as though you are experiencing a life-threatening situation. Yes, it is only a math exam, but your body is reacting as though a mass murderer broke into your home and you must either fight for your life or run, or you are walking down the street and come face-to-face with a grizzly bear. This is why your mind goes blank. If you faced a grizzly bear, would reciting the quadratic formula be anywhere close to your top priority?

How can you combat this crippling panic. You can utilize a strategy entitled "the data drop off." This is something you can use during your exam to combat test anxiety. When does the extreme panic set in to the point where you forget everything? When you first look at the exam? So, how can you battle that?

The data drop off (or memory dump) consists of a short exercise where the student receives his or her exam but does not look at it right away. Instead, on a separate piece of paper, the student immediately takes a few minutes to write down various formulas or algorithms. This way you can write down what you need to remember for the exam before you have a chance to forget it. Then, this information is there for you on your data drop off when you need it. The tables below show examples of a data drop off for various exams. Table 7.2 represents a sample data drop off from introductory algebra. Table 7.3 represents a sample data drop off from college algebra. Table 7.4 exemplifies a sample data drop off from basic statistics. Again, keep in mind this may not make sense right now, and it may look like complete gibberish before taking the related math class, but these are formulas or algorithms that students must remember and ways to remember them. The left side indicates what the student is attempting to recall, and the right side focuses on the formula or the way to remember the content.

You should use the data drop off when you receive your exam; however, you should start practicing the data drop off a few days before the exam. How I do I study for a math exam? Well, here is one way. Practicing the data drop off will not only prepare you for the exam day, but it is a great study strategy. You should have at least four or five rehearsals where you practice and practice your data drop off.

TABLE 7.2

Sample Memory Dumps from Pre-Algebra/Introduction to Algebra

Topic	How to Remember it
Order of operations	PEMDAS or Please Excuse my Dear Aunt Sally
Equation of a line	$y = mx + b$
Find slope	$\dfrac{y_2 - y_1}{x_2 - x_1}$
Steps to solving equations with fractions	PFCVAD (Parenthesis, Fractions, Combine like terms, Variables to one side, Add or Subtract, Divide)
Laws of exponents	$x^2 x^6 = x^8$ Product rule—add powers
	$\dfrac{8x^9}{4x^4} = 2x^5$ Quotient rule—divide the numbers, subtract the powers
	$(3x^9)^2 = 9x^{18}$ Power to power rule—raise the number to the power and multiply the powers
When you evaluate, use parentheses! Evaluate $4x^2$ when $x = -3$	$4(-3)^2$
Perimeter of rectangle, which is all around the rectangle	$P = 2L + 2W$
Area of a rectangle, which is inside the rectangle	$A = L \cdot W$

TABLE 7.3

Sample Data Drop Off from Intermediate or College Algebra

Topic	How to Remember it
Quadratic formula	$x = \dfrac{-b \pm \sqrt{b^2 - 4ac}}{2a}$
Domain	All the values that x is allowed to be
Range	All the values that y is allowed to be
Exponential equation	$y = ab^t$
Vertical asymptote	$y = \dfrac{3x}{x - 8}$ Set the denominator to zero and solve for x $x - 8 = 0$
Horizontal asymptote	1) When the highest power is in both the top and bottom, you divide the numbers in front of the highest power. So $y = \dfrac{4x - 1}{2x + 3}$ The answer is $y = 2$ 2) When the highest power is on the bottom, the answer is always 0. So if we have $y = \dfrac{4x}{x^2 + 1}$, since the x^2 is on the bottom, the answer is $y = 0$.

TABLE 7.4

Sample Data Drop Off for Introductory Statistics or Quantitative Reasoning

Topic	How to Remember it
Measures of central tendency	Mean, median, and mode
What are mean, median, mode, and range	Mean is you add all the numbers and divide by the amount of numbers you have. Median is you list all the numbers from lowest to highest and pick the middle number. Mode is the number that appears most. Range is the highest minus the lowest
Histogram	Use a histogram when there is a large range and don't forget the intervals have to be even
Dot plot	Use a dot plot when the range is small; don't forget to list every number, even if it isn't a value
Box plot	Remember the five-summary data are the lowest value, the first median, the median, the second median, and the highest value
More on box plots	Remember, 25% of the data are between the lowest value and first median; 50% are between the first median and second median; the highest 25% are between the second median and highest value
Probability with "and"	$P(A) \cdot P(B)$
Probability with "or"	Can this happen together? Yes, then, $P(A) + P(B) - P(A \cdot B)$. No, then, just $P(A) + P(B)$
Probability—"without replacement"	Be careful of the second fraction. The denominator will go down by 1 if you pick 1 and do not replace it
Symbols!	μ = mean; σ = standard deviation
Empirical rule	See below!

99.7%

95%

68%

0.15% 2.35% 13.5% 34% 34% 13.5% 2.35% 0.15%

$u-3\sigma$ $u-2\sigma$ $u-1\sigma$ u $u+1\sigma$ $u+2\sigma$ $u+3\sigma$

Timed Tests!

One of the major contributors to test anxiety, for students, is the dreaded timed test.

> *I would get my exam, and I could just hear the clock ticking really loud. Sometimes I couldn't even focus or read the problem, because I was so worried about running out of time.*
>
> George

We could have an endless debate as to whether timed tests are fair. Does a timed test limit the true knowledge that a student could demonstrate? More specifically, if a student had more time on an exam, would that lead to a score that is more reflective of his or her knowledge? It is possible, but timed tests are a reality. As mentioned earlier, you will likely need to pass an exam to become certified or enter your career field, and this test will likely be timed.

How can you conquer a timed test? You need to prepare and practice. Politely ask your professor how many questions will be on the exam. You could phrase the question as:

> Professor (name), could you let us know how many questions will be on the exam? I just want to prepare as best as I can.

Some professors may provide you with an exact number of questions for the exam; some may provide you with a ballpark number, and some may refuse to provide any number. Next, you will want to determine how much time you will have for the exam, and most, if not all, professors will provide this information in advance. Will you have the full class to complete the test, or just part of the class? Sometimes if class meets for a lengthy time (e.g., over 2 hours), the professor may allot part of the class for the exam and cover new material during the other part.

How can you prepare and practice? As you complete your homework assignments or the study guide, try and time yourself. How long does it take you to complete five or 10 problems? Additionally, how long does it take you to complete both computational problems and word problems? Next, try and complete a certain amount of math problems within your allotted time for the exam. It certainly helps to know exactly how many problems are on your exam, but practicing problems within a time frame will help you prepare for the timed exam. Ultimately, however, you will want a game plan to hit the ground running and use your time efficiently. When attempting a math problem, ask yourself two questions: What is the eventual goal? What is the first thing I do? Consider this problem from basic algebra:

Find the equation of a line that passes through the points (2, 3) and (5, −7).

Keep in mind the following steps will not make sense until you have mastered the content for this type of problem, but hopefully the game plan and transitions make sense. The goal of this problem is to find the equation of the line. That means we need to arrive at what is referred to as $y = mx + b$ form. And "equation of the line transitioning to $y = mx + b$" should be in your data drop off. Ultimately, we need to find the slope, which is "m" and the y-intercept, which is "b." Which do we find first?

Step 1: We need to find the slope first, and we do that using the formula, $\dfrac{y_2 - y_1}{x_2 - x_1}$ (something else in your data drop off!). Where do we find those values? We go back to the two points that were given earlier, $x_1 = 2$, $y_1 = 3$, $x_2 = 5$, $y_2 = -7$. So, $\dfrac{-7-3}{5-2}$, which is $-\dfrac{10}{3}$, and that is the slope.

Step 2: How do we find the y-intercept, which ultimately will give us our goal of getting this into $y = mx + b$ form. We use another formula, known as the point slope formula, but we need the slope within that formula, which is $y - y_1 = m(x - x_1)$. We already know that $x_1 = 2$, $y_1 = 3$. We have learned that "m," the slope, is $-\dfrac{10}{3}$. Now, we plug those values into the point slope form, and we solve for y: $y - 3 = -\dfrac{10}{3}x + \dfrac{20}{3}$. We solve for y by adding 3 to both sides. This gives us our equation of the line. When, we add 3 to both sides, we arrive at our answer. $y = -\dfrac{10}{3}x + \dfrac{29}{3}$.

Additionally, being as prepared as possible, in terms of practicing and studying the content, will also allow you to be more efficient with your time.

The Day of the Exam

We have outlined how to prepare for your exam. Now, how can you handle the day of your exam and the exam itself?

Try to Get Enough Sleep

Being a college student can be brutal. Between classes, studying, working, and possibly taking care of family, sometimes sleep gets the shaft. However, it is imperative that you get enough rest before your exam. Plan to get at least seven or eight hours of sleep the night before your exam. Do not sacrifice any sleep the night before the exam to cram or study. That will simply lead to more anxiety and fatigue.

> *I used to stay up most of the night [before an exam] trying to cram for my test. First of all, I would do that because I hadn't studied enough before, so I was getting stuck on things. The more I would get stuck and make mistakes, the more frustrated and nervous I would get. Then, I would just get more and more tired. Then, I was in even worse shape for my test.*

Derek

Self-efficacy refers to an individual's belief in his or her ability to act in the ways necessary to reach his or her goals. More specifically, self-efficacy refers to your own belief that you can achieve your goals, and your own belief that you have the skills necessary to achieve your goals. However, when an individual feels tired or fatigued, his or her self-efficacy is lower. Basically, when you are tired, you do not feel as confident that you can reach your goal.

Eat Well During the Day

Keep in mind that your body is a machine, and it runs on the food you provide. More specifically, your body converts food into energy, and this process is called metabolism. We need this energy to move, think, and react. However, you will want to eat healthy food, so that your body works efficiently.

If your math exam in the morning, eat a nutritious breakfast. If your exam is later in the day, be sure to eat regularly throughout the day. You will want to avoid processed foods or foods with refined sugars. Why? Because such foods can lead to a person feeling fatigued, which as discussed earlier, can lead to a lower self-efficacy. If possible, you will want to consume fruits, vegetables, or lean protein.

I get it. Being in college often leads to consumption of fast food. While I cannot imagine doing this now, throughout college, I am sure I qualified for the Guinness Book of World Records for most consecutive days for eating at McDonalds, Burger King, or Wendy's. Yes, I was regular at all those restaurants between classes. Basically, fast food was in line with my budget and worked within my busy schedule. However, consider, at least on exam days, eating healthier. For example, instead of a big mac and fries at McDonalds, how about a grilled chicken sandwich with some apple slices?

How about coffee? After all, coffee can provide that jolt of energy that many of us need. Even now, as a professor, on my long days of teaching and then a workout, I will consume two to three cups of coffee throughout the day. Be careful; while coffee can provide energy, it also contains large amounts of caffeine, which can dehydrate you, and dehydrations leads to fatigue. What is the solution? Consume a bottle of water for every cup of coffee that you consume. This will limit the dehydration. In fact, you will you want to drink water, in general, to stay hydrated.

Try to Stay in a Good Mood

Returning to the concept of self-efficacy, being in a better mood or state of mind can also increase your self-efficacy. Of course, we cannot control external factors. How many times have you woken up in a good mood only to have various external factors of life events bring you down? However, there are actions we can take to put ourselves in a better mood. How about listening to some of your favorite songs? In fact, right now, list your five favorite songs that put you in a good mood! Studies show hearing our favorite songs can improve our mood. How about having breakfast or lunch with a family member or friend who has a positive impact on you? If time permits, how about a workout of some sort? This could even include a brisk walk around campus. Exercise creates endorphins, which have been proven to lower stress, enhance mood, and improve a person's well-being.

Let me relay a personal anecdote of an event that may or may not have led to my failing of a final exam. It was Final's Week of the fall semester of my senior year in college. My exam, for a high-level math class, was scheduled for 5:30 pm that evening. I had been studying for that exam along with several others. Before this exam, however, I had a Christmas work party to attend. Again, I was a student tutor at the Academic Support Center at my former community college. At that point, I had worked there for over 2 years and knew everyone well. I met most of the people at the Academic Support Center when I started college. I started working there as a tutor after my freshmen year. Those people felt like family to me. However, I left the party feeling very hurt. I had worked there regularly, and I was about to take a semester off to focus on student teaching. I was anxious about the upcoming experience and felt sad about leaving, even just temporarily. However, none of my superiors acknowledged that I would be leaving. Did I expect the party to be for me? No. Did I expect a major send off? No, but I thought I would get an acknowledgment. I had worked closely with these people who were like family to me, and I felt hurt and unappreciated.

Not only did I leave the party feeling upset, but I lost track of time, and I was late to my math exam. On top of my hurt feelings, I felt frazzled and rushed as well. While I had prepared for the exam, I found myself drawing several blanks. I earned a 48% on the exam and a "C–" in the class. Had I scored any lower, I would have needed to repeat the class to graduate.

You be the judge. Put yourself in my position. What should I have done differently? Or what would you have done differently? Would you have not attended the party? Would you have been able to understand that while this outcome was upsetting, the exam was more important? Is it possible that I did not study enough and that contributed to my failure? Think about potential situations that could arise for you that lower your spirits or self-efficacy before your math exam. How can you prepare yourself mentally to keep your exam as a high priority? The bottom line is we must be able to contend with life's events.

Online Test-Takers

The aforementioned suggestions should apply to both in-person and online test-takers. Online test-takers may have more flexibility in that you may not need to take your exam at a very specific time. You may have a window of time (e.g., Wednesday– Sunday) to take your exam. However, you will likely need to complete your exam in one sitting (you cannot sign off and back on), and there will still be a time limit. Of course, you may also be taking a virtual class, which requires you to sign on at a very specific time and take your exam with the class with your professor watching.

Online test-takers have additional responsibilities. You will need to ensure that your Internet connection is strong and reliable. You will want to practice the steps needed to access your exam. Additionally, even if you can complete your exam asynchronously, you will likely still need a functioning webcam. To ensure there is no cheating, your professor will still record you taking the exam. Therefore, make sure that your webcam is operating correctly. Additionally, do your best to ensure there are no other technical issues. Ask your professor if there is a way to test out any equipment, such as Respondus, that you need before the exam. Do not allow technical issues to thwart your success.

You Get the Test!

The moment has arrived. Your professor passes out the exams. Or you sign on to take your exam online. If you are taking the exam in person, learn from my mistake (mentioned earlier) about arriving late. Ensure that you arrive to the exam in enough time and that you have the needed supplies (e.g., calculator, scratch paper, pencils).

Now, Use the Data Drop Off (Memory Dump)

Whether you are physically receiving the exam or opening the exam online, this, as mentioned earlier, is where the panic occurs. Therefore, this is where you will want to utilize the data drop off (memory dump), which is discussed earlier. In fact, you will want to allot the first five to 10 minutes to complete the data drop off on scratch paper. Keep in mind, again, you should have been practicing the data drop off before the exam. This way, the formulas or samples will be there when you need them.

Start Where You Feel Most Comfortable!

It is time to start the exam but let me first pose a question. Let us say that you are anxious about attending a work party, a wedding, or some kind of social engagement, because there is someone there (or multiple people) who you do not want to see for various reasons. We have all been there! When you apprehensively attend this function, do you walk right over to the person or people whom you do not like or do not want to see? Of course not. You find someone with whom you feel comfortable with and talk with them. What happens? You start to feel more comfortable and confident. Why not apply the same tactic to an exam? When you open your exam, instead of attempting the first question, search for the question in which you feel most comfortable. Rather than question # 1, that could be question # 4, 6, 10, or 15! When you answer a question in which you are most confident in your answer, your confidence and self-efficacy will increase. This contrasts with starting off with a question in which you may struggle. What happens then? Your anxiety will increase, and your confidence will decrease. Therefore, keep searching for questions in which you are most confident. The more questions that you know you are answering correctly, the more confident you will feel. If you get stuck on a problem, leave it be and return to it later.

Seemingly Contradictory Advice

Math teachers often give students confusing advice for exams: "Check over your work but go with your first instinct and do not second-guess yourself." That sounds contradictory, right? As a student you could justifiably question, "Do you want me to check over my work or just leave it alone and trust my judgment? Make up your mind!" "Checking over your work" refers to checking for simple mistakes. Did you make any simple arithmetic or multiplication mistakes? For example, did you mistakenly write that $2 + 3 = 6$? Did you accidentally miscopy something from the test to your scratch paper or vice versa? Walk through each question again to check for simple mistakes. So, what about "go with your first instinct." When it comes to the exam, you either know the content, or you do not. More specifically, you either know the formulas and process for a problem, or you do not. As an example, you either know the quadratic formula or you do not. An exam is not the time to second-guess yourself.

Pacing the Exam

Let us put this all together. Spend the first 10 minutes completing the memory dump. Spend the last 10 or 15 minutes checking for simple mistakes and that you completed the requirements of each question. Use every minute of your test time!

One of the major distractions for in-person test-takers is students finishing and leaving the room.

> *That used to drive me crazy and make me freak out. When other kids would get up, turn in their tests, and leave, I thought they were finished because they did so much better than me. It made me think I was going to fail.*
>
> Alicia

I realize this can be difficult, but you must block out other students finishing and leaving before you. The reality is that several students who submit their exams early do not excel or even pass their exams. More specifically, oftentimes, these students omit questions or partially answer questions. Quite frankly, as a teacher, I have become so frustrated with students not using the allotted time to carefully answer questions, I have noted the time a student spent on the exam. For example, if a student had an hour for the exam but only spent 30 minutes on the exam and made careless mistakes, I have noted "only 30 minutes used!" on top of the exam.

The Waiting is the Hardest Part

Waiting for the results of an exam can be brutal. You may find yourself going over the exam in your mind, questioning if you answered various questions correctly and second-guessing yourself. Keep in mind that we have all been there. I wish I had some advice. The comprehensive exams for my doctorate entailed completing three closed book essays each day for a three-day period. I then had to wait two weeks for the results, and I nearly lost my mind. Whether your class is in-person or online, your professor will likely post your exam results online. Pay attention to when this information may be posted.

Celebrate!

In the next chapter, we will discuss assessing your exam results and where to go from there. However, give yourself something to look forward to after your exam. How about dinner at your favorite restaurant? How about some ice cream or something decadent? You could see a great movie with your significant other or a friend. If you have children, how about a family night out? Or you could just make time for yourself to binge watch some of your favorite shows! A positive end to test-day can make the day less daunting.

Why in the World? Part 7

Why in the World Can't I just Divide Stuff to Simplify?

Let us say we have the following expression or possibly we have done previous work to get to this expression: $\dfrac{x^2 - 4x - 32}{x + 4}$.

We hear that mathematical expressions should always be simplified. So, to simplify, why can't we divide $\dfrac{x^2}{x}$ and get x; divide $\dfrac{-4x}{x}$ and get -4; divide $\dfrac{-4x}{4}$ and get $-x$; divide $\dfrac{-32}{4}$ and get -8? This would make our answer: $x - 4 - x - 8$. And hey, we could simplify to -12. Right? Wrong!

Why this makes no sense: Let us look at a similar example that could lead a student to thinking the above work is correct. How about? $\dfrac{4x^2}{4x}$. In this problem, we divide $\dfrac{4}{4}$ to get 1; we divide $\dfrac{x^2}{x}$ to get x. Therefore, the answer is $1x$ or x. Why can we divide out in the second expression but not the first?

Let us try to understand: When quantities are attached by multiplication (second problem), we can divide what can be divided, whether that includes numbers or exponents. However, when quantities are attached by addition or subtraction, we cannot divide, to simplify, unless one term divides into all terms. For example, notice how in the first problem, I divided the x into x^2 and $-4x$ but not -32? I divided the 4 into $-4x$ and -32 but not x^2.

So, can we ever divide across addition or subtraction? Yes. Consider this expression: $\dfrac{x + 4x}{x}$. I can divide $\dfrac{x}{x}$, which gives me 1; I can divide $\dfrac{4x}{x}$, which gives me 4. Therefore, my answer is $1 + 4$, which is 5. Again, I was able to do this because I could divide the x into every term.

This may shed some light on these problems, but this still seems like some arbitrary rule that some mathematician made up, right? No, the rule can be tied to arithmetic. Consider this expression: $\dfrac{3 + 5}{3}$. We know the correct answer is $\dfrac{8}{3}$. However, to try and simplify earlier, could we have divided $\dfrac{3}{3}$, which is 1, then add 5? That would give us 6, which we know is wrong. But how about this: $\dfrac{3 + 9}{3}$. We know the answer is $\dfrac{12}{3}$, which simplifies to 4. Again, to try and simplify earlier, we could have divided $\dfrac{3}{3}$, which is 1; then divide $\dfrac{9}{3}$, which is 3. And then $1 + 3 = 4$. That worked because we could divide the 3 into all terms.

But wait! We never answered the original question. How would you simplify $\dfrac{x^2 - 4x - 32}{x + 4}$? We cannot add or subtract across the numerator or denominator. We just discussed why we cannot divide out. However, there is another form of simplification, and that is factoring. We can factor the numerator to $\dfrac{(x-8)(x+4)}{x+4}$. This is because 8 and 4 multiply to 32, and $-8+4=-4$. Can we simplify further? Yes! We can divide $\dfrac{x+4}{x+4}$, which is 1. That leaves with $x-8$ as the answer. Why can we divide out $\dfrac{x+4}{x+4}$? Because the $(x+4)$ is attached to the $(x-8)$ by multiplication.

Activity

- Think about poor experiences on previous math exams. How did this make you feel about yourself? Remind yourself that those exams are in the past. No one is looking at those exams and making judgments about you. Remember, college is a time to start fresh.

- Reflect on a specific time where you experienced math test anxiety. Do you remember when the panic set in? Do you feel the data drop off (memory dump) would have helped you?

- Even if you have yet to attempt a math class, I want you to try the data drop off. Find a book that you like and pretend that there is an exam on one of the chapters. In fact, design some questions for a test. Next, construct a data drop off to help you remember the information.

- If you have already attempted a college math class, review an exam. Can you construct a data drop off that may have helped you?

References

Adelman, H., & Taylor, L. (2010). *Mental health in schools: Engaging learners, preventing problems, and improving schools*. Corwin Press.

Goldstein, D. S. (2010). Adrenal responses to stress. *Cellular and Molecular Neurobiology*, 30(8), 1433–1440.

Seemiller, C., & Grace, M. (2016). *Generation Z goes to college*. Jossey-Bass.

8

Learning From Success, Failure, and Experience

The wait is over! You can view your exam results and even your exam. How do you assess your progress? More importantly, where do you go from here? We will address those questions, and more, in this chapter.

Assessing Your Exam

What are the possible reactions when you view your grade and your exam? You had a good feeling, but you still rejoice when you see that you excelled; you are in disbelief that you did so well, and you jump for joy in shock; you feel disappointed in your results; you thought you excelled, or at least passed, and are shocked that you failed, and finally, you knew you failed, and seeing the grade only provides confirmation. Chances are you will experience one of those reactions when you see your exam result.

Check Out Your Mistakes

Regardless of your grade, and we will get to that, what kind of mistakes did you make on the exam?

> *I would get my [math] exams back, and I would want to punch myself. I totally understood the mistakes I made. It was silly and stupid little things like I would use the wrong formula or the wrong procedure.*
>
> Kara

> *It was really stupid things like I would divide or multiply wrong, or I would copy things down wrong, or just other mistakes where I just figured it out when I saw the right way to do it. Really frustrating.*
>
> Bob

DOI: 10.1201/9781003614142-8

Of course, there are other types of mistakes.

> *When I got my [math] exam back, I was as lost as when I took it. Even when I saw the answers or the way to go about the problem, I was still so confused.*
>
> Alicia

> *Getting the exam back didn't help me at all. I was still so confused. Even when my professor tried to explain my mistakes to me and how to do the problem, I still didn't get it. Sometimes I even thought I knew what I was doing, but when I got my test back, I realized I was totally clueless.*
>
> David

As illustrated above, there are two types of mistakes:

Mistake (1): This type of mistake occurs when where a student can quickly or even eventually understand what went wrong and be able to learn from it and go from there.

In Figure 8.1 is an example:

In the example in Figure 8.1, the student made a mistake when factoring, but it seems as if he or she understands the process of solving a quadratic equation. The student correctly set the equation equal to zero. The student seems to have a general understanding of factoring, but he or she needs work and practice in factoring. This is important, as factoring is imperative, not only to solving quadratic equations, but in the STEM pathway. Ultimately, the student is on the right track. Figure 8.2 provides another example for Mistake 1.

Solve: $2x^2 - x = 15$ (a quadratic equation)

Wrong answer:

$(x + 5)(2x - 3) = 0$ After factoring

$x + 5 = 0, 2x - 3 = 0$ Set each binomial to zero:

$x = -5, x = \dfrac{3}{2}$ This is after solving for the variable.

What went wrong:

The student made a slight error when factoring. The correct binomials, after factoring, should have been: $(2x + 5)(x - 3) = 0$

The correct answer, therefore, is $x = -\dfrac{5}{2}, x = 3$

FIGURE 8.1

Mistake 1—An example with a quadratic equation.

Solve the following system of equations by elimination:

$$\begin{cases} 2x - y = 12 \\ x - 2y = 48 \end{cases}$$

What the student did: To eliminate the y variable, the student multiplied the top equation by 2 and the bottom equation by –1.

$$\begin{cases} 2(2x - y = 12) \\ -1(x - 2y = 48) \end{cases}$$

$$\begin{cases} 4x - 2y = 24 \\ -x + 2y = 48 \end{cases}$$

However, note the signed number error in the second equation: (–1)(48) = –48, not 48.

This leads to an incorrect answer for the x variable:

$$3x = 72$$

$\dfrac{3}{3}x = \dfrac{72}{3}$, which simplifies to $x = 24$, which is incorrect, and when the student substitutes this value into the y variable, both values will be incorrect.

This is a minor error. Clearly, the student understands how to solve a system of equations. In fact, it seems as though the student understands the rules of signed numbers; however, he or she simply made a small error with multiplying signed numbers one time.

Let us go back and solve this the correct way:

$$\begin{cases} 2(2x - y = 12) \\ -1(x - 2y = 48) \end{cases}$$

$$\begin{cases} 4x - 2y = 24 \\ -x + 2y = -48 \end{cases}$$

$$3x = -24$$

$$x = -8$$

FIGURE 8.2
Mistake 1—An example with a system of equations.

This will allow the correct answer for the y-variable. We can substitute -8, for x, into the second equation.

$-8 - 2y = 48$

$-8 - 2y = 48$

$+8 \qquad +8$

$-2y = 56$ and $y = -28$

Therefore, $x = -8$ and $y = -28$

FIGURE 8.2   (Continued)

Solve: $2x^2 - x = 15$ (a quadratic equation)
Wrong answer:

$2(x^2 - x) = 15$

$x = \dfrac{15}{2}, \; x = 2$

FIGURE 8.3
Mistake 2—An example with a quadratic equation.

Mistake 2: This type of mistake demonstrates that the student lacks the basic understanding of the problem and likely the prerequisite skills. The student can view the correct answer, read the correct process, and even get help, and the original question makes no sense. We will revisit the previous two problems but with more severe mistakes. See Figure 8.3 for the first example. Reviewing the work from Figure 8.3, we can see the student is completely off track with his or her work. The student does not seem to understand what a quadratic equation is, much less how to go about solving one. He or she certainly does not understand factoring. This student is now at a major disadvantage because he or she needs to master a previous concept from a previous unit while trying to master new concepts from a new unit. Too many of these types of errors (Mistake 2) create serious doubt about the success in the class.

Let us also revisit the system of equations problem (Figure 8.2) and examine another example of Mistake 2 (see Figure 8.4).

Solve the following system of equations by elimination:

$$\begin{cases} 2x - y = 12 \\ x - 2y = 48 \end{cases}$$

Here is what the student did:

Answer: $3x - 3y = 60$

The student simply added vertically. Clearly, the student understands the rules for adding signed numbers; however, he or she has no idea how to go about solving a system of equations. Moreover, it is concerning that the student would think that is an acceptable answer. After all, the student should have a specific value for both x and y. That is how we solve equations. Therefore, the student may lack an understanding of basic equations as well.

FIGURE 8.4
Mistake 2—An example with a system of equations.

The worst course of action a student can take is to not even view the exam and study the mistakes. Most professors will allow you to keep your exams. Some professors may allow you to view but not keep the exam, as these professors may reuse the exam and do not want the exam to circulate. If this is the case, you may want to make an appointment during your professor's office hours to get a better look at the exam.

Viewing a past exam can be especially difficult for online students.

> *When I took my [online] math class, I could never figure out how to see my exam and what questions I got right or wrong. I guess I didn't ask the right questions or make enough effort to find out. So, when I had to take the class over, I made sure to ask my instructor how to look at my exam, and the next time I could see the exam and my professor's comments about what questions I got wrong.*
>
> Kara

Kara's issue viewing her exam can be common for online students. The reality is that the technology is more complex. Before the exam, be sure you know how to access your exam with the instructor's feedback. If you have difficulty, contact your institution's technical support. This is very important. Students who do not view their mistakes put themselves on the path to failure.

TABLE 8.1

Evaluating Your Grade

Grade	Recommendations
A (90–100%)	This is an amazing accomplishment! An "A" indicates that you clearly understand the content and are in great shape moving forward. You will still, however, want to review any mistakes
B (80–89%)	A "B" is a very good grade. You will want to review your errors. If you had "Mistake 1" errors, clean up any problems. If there were any "Mistake 2" errors, be sure that you meet with your professor to ensure that you master the content
C (70–79%)	It is important to keep in mind that as you progress in mathematical content, the material you just learned will eventually serve as prerequisite material. If you received a "C," you passed, but you made several mistakes. If you do not review such mistakes, and ensure that you understand the prior content, you could set yourself up for future struggles
D (60–69%)	A "D" may be passing, but it is a lousy grade. Most colleges do not even accept "D" grades when students transfer. Like the "C" grade, and even more so, you need to review and address your mistakes
F (0–59%)	As you can see from the percentage range, an "F" is very broad. Failing an exam is alarming, but by how much did you fail, and what kind of mistakes did you have? If your grade was in the 50s and you made mostly "Mistake 1" errors, you might be alright going forward if you address these mistakes and ensure you understand the content. However, if you received a grade of 10%, 20%, or 30%, and worse off, you made mostly "Mistake 2" errors, you are in serious trouble. First, grades such as 10%, 20%, 30% will make it difficult to pass the class; it will be very challenging to raise your final average to the point of passing. Secondly, you likely need to learn most of the content from the previous unit while attempting to apply this content to the new content and mastering the new content as well

How About your Grade?

See Table 8.1 for the recommendations for each grade.

Use Your Exam as a Study Guide

Just as you should focus on missed questions in your homework assignments or classwork, you should use your exam as a study guide for future exams, especially the final exam. In general, most final exams are comprehensive in that they cover content from the entire term. As you prepare for your final, review mistakes from unit exams. But do not take my word for it.

> *I started doing so much better on my final exams once I started going over my old exams. I don't know what I had been thinking in the past. Maybe I figured that mistakes I made in the past would just go away.*
>
> Emily

My advice. Save all your exams and go over what you got wrong. It will come up again, on the final!

Derek

I used to stress out so much studying for my final exams [in math]. I didn't know where to start. I would look at the practice test [for the final] and freak out, because I couldn't remember how to do so much stuff, but it helps so much to go over your old exams right away and correct your mistakes and understand what you did wrong.

Bob

Learning from Success

You passed your math class! Congratulations! This is a major accomplishment. However, if you need to complete more math classes for your degree, you need to reflect on your experience. After all, math is linear and progressive, and there is a good chance that you will need to apply the content from your previous math class to the next class. Here are some questions to help you reflect:

- Did you receive a "C" or a "D"? If so, keep in mind that you have gaps, and you need to focus on those gaps.
- Review for your next math class. What content from your previous class do you need for the next math class, and do you have a good grasp on that content?
- Did you devote much time to your class? Or were you a "coaster"? More specifically, did you pass your class without maximum effort? This may seem like an antagonistic question. However, sometimes a student's first math class is not particularly challenging, possibly because the student encountered a lot of review material. Andrea reflected on her first class, which was introduction to algebra.

I went to class and did the work, but most of the stuff I already knew. I guess I remembered some algebra from high school, so I didn't really study much. I guess I did the least amount of work I needed to.

Andrea

Coasting can catch up to students, as math classes get progressively more difficult and familiar content may wane. If you were a "coaster," you need to be prepared to put more time into your next math class.

Should I Withdraw from the Class?

On their academic calendar, colleges list important dates throughout the term, and one of those is the withdrawal date. This is the last day to withdraw from a class. The exact date varies, but most colleges set the withdrawal date about two-thirds of the way through the term. Students who wait any longer will need to either stay in the class or accept a failing grade. Generally, students who withdraw by this date will not receive any kind of refund and will receive a grade of "W" on their academic transcripts. The conversation focusing on withdrawing is uncomfortable and unpleasant; however, at times, it is necessary.

Withdrawing from any class should be a last resort. After all, you have invested time and money into a class and are essentially getting nothing in return. Therefore, when is it time to withdraw? The best answer is when there is no realistic hope of passing the class. More specifically, if you are in a position where you are failing and even if you received 100% on the remaining exams, you would still fail, it is time to withdraw. Keep in mind that college is not high school. I have spoken with many students and even former high school teachers who discussed how difficult it is to fail high school. Several students have relayed that when they were in danger of failing a class, and especially not graduating, they were given "extra credit assignments" by their teachers. After completing these assignments, the students would somehow pass their classes. Former high school teachers have relayed that when a student was failing, and in danger of not graduating, administrators and others pressured them into finding a way to pass the students. In fact, the so-called "extra credit assignments" were merely busy work, a justification for passing them. Students have even mentioned that they were so used to passing in high school that they were even in denial about failing a college math class.

> *I just assumed I would pass. I always passed. I figured something would happen to get me to pass my [math] class.*
>
> Andrea

> *I kept asking my professor if I could just have an extra credit project, since I was failing. I couldn't understand why she wouldn't just give me one.*
>
> Peter

Before withdrawing, you should have a conversation with your instructor and ask if he or she recommends that you drop the class. Oftentimes, professors will contact students who need to withdraw. Withdrawing from a class, just like failing a class, can be a deflating feeling and can make a person feel like a failure. Failure is never an easy conversation, but it is something that we must address, and we will do so in the next section.

The Reality of Failure

As a teenager and young adult in the 1990s, one of my favorite television shows was Beverly Hills 90210, which followed a group of young people through high school, college, and the real world. In the episode where the group graduated from college, part of their president's commencement speech was,

> *For four years, we have challenged you; we have graded you; we have prodded you to succeed. About the only thing we haven't done is talk to you about failure. That's too bad. Not just because failure is a part of life, but because the surest way to succeed is allow yourself the freedom to fail.*

Failure is indeed a part of life. I am an avid professional baseball fan, and part of my fascination with the game is that even the best hitters fail (make an out) at least seven out of every 10 times at the plate. Pitchers have bad outings and must rebound. This is because the major league baseball season is so long that failure at some point is inevitable. As humans, we exist for 365 days (sometimes 366) each year for many years. Our time on this planet will involve success and failure. Failure still hurts, and as I mentioned earlier, the fear of failure is a major cause for anxiety. Most importantly, please realize that when it comes to failure, you are not alone.

> *When I realized, I wasn't going to pass my first math class, I just cried. I kept thinking, "I'm a failure. I'll never succeed."*
>
> Emily

> *It almost felt like a worst nightmare come true. I hated math, and now I failed a college math class.*
>
> Rosemary

> *I just wanted to quit. I never wanted to come back to school. It just felt impossible. I was never going to pass a math class.*
>
> Dan

It is noteworthy that all these students eventually passed their required math classes. In this section, we will discuss paths to take if your attempt at a math class is unsuccessful. However, before that, I would like to share one of my own experiences with failure.

My Experience with Failure

I will admit that I did not flunk out of college or even fail any of my college classes; however, I did worse than that. I failed my first year after college.

More specifically, I failed my first year in the "real world." I consider that worse than failing any college class, because isn't college an investment for the real world?

I graduated with a bachelor's degree in math and a minor in elementary education. I was excited about entering the "real world" and earning money and living in my own apartment. I also picked up a math class to teach at the community college where I attended and had been working as a tutor. This was especially exciting, as I knew my eventual goal was to be a math professor at a community college someday. However, I had no idea how hard the real world would be.

I received an offer to teach sixth grade math and science in a middle school in the New York City Public School system. I entered with enthusiasm but was met with frustration and exasperation. My classes were out of control from Day 1. Any of the "classroom management" techniques that I learned in college failed. I cannot even say that I ever taught any of my students in that time. My classes were overcrowded, and classroom resources were in short supply. To make matters worse, I did not even have my own classroom. I was a "floater," which meant each class was in a different classroom almost each day. I spent my class periods yelling at my students to "sit down and be quiet." I was miserable all the time. I dreaded going to work and was often in tears on my drive there.

It took eight days for me to reach my breaking point. After admonishing one class for the umpteenth time and threatening detention, the students started mouthing back, even worse than before. When the bell rang, I grabbed my belongings, and on my way out, I mouthed, "fu----- kids." Only I did not say "fudge." Yes, I said what you are thinking I had said. As luck, would have it, I had the exact same class the next period, as I was assigned to teach both math and science. As soon as class started, one of the students said, "We know you said fu----- kids, and if you don't take away the detention, we are going to the assistant principal (AP)." I refused, and a few of them left the room to report me to the AP. I was so angry and exhausted that I did not care. Several minutes went by. What happened? While I sat feeling hopeless, the students sat around and talked; some played cards; some even meandered around the room. The AP eventually entered and scolded the class; however, he quietly said to me, "I'd like to see you later." Was I afraid? Not really, I just did not care.

Did I get fired? No, the AP reprimanded me and informed me that such behavior could bring charges against me. However, I wanted to get as far away from there as possible. So, I contacted my former supervisor at my old job at the tutorial center and explained my situation and asked if there was any work for me. Not only was she able to give me some hours at the tutorial center, but she mentioned that the math department chair was looking to fill a late-start math class at the community college. Basically, I got my old job back as a tutor, but now, I was teaching two college math classes. I abruptly quit my middle school teaching position. Everything was coming up roses, or so I thought.

In college, I thought that working as a tutor on an hourly wage and teaching as an adjunct (part-time) would be a great job. However, I learned very quickly that I needed a full-time job with health benefits to survive in the real world. Worse off, I was in a terrible relationship and made the horrendous decision to get engaged. Please do not ask me why. I still do not know! This relationship led me to feel depressed, anxious, and exhausted, and I started sleeping too much and even missing work.

One of my major excitements when entering the real world was my own apartment. A very nice, retired couple converted the basement of their home into an apartment, and I chose this apartment. Besides paying my rent on time, all they asked was that I care for the apartment. Because I was dealing with so much, anxiety, depression, and fatigue, my housekeeping slipped and slipped badly. My landlords must have suspected as much, because one weekend when I was out of town, they entered my apartment and were in a state of shock. They found a sink overflowing with dirty dishes, a tray of brownies rotting away, the thermostat turned all the way up, a dirty and wet carpet, and a bathroom that would have shocked modern science! My landlords sent me a registered letter of eviction. Think about it. My depression and anxiety became so bad that I had difficulty expending energy for basic housekeeping. So, if mental health has ever hindered your success for something as complex as a math class, do not feel bad!

I enjoyed teaching math part-time, but because of my anxiety and depression, I missed so much work at the tutorial center that I was reprimanded and then unceremoniously dismissed. Again, these were the people who gave me a chance after my failure as a middle school teacher, and I thanked them by blowing off my responsibilities. This led to another job search, and I did obtain a full-time job as an academic advisor for a small college. The job was appealing. I would be working with college students and guiding them in their academic plan. But wait, I even managed to screw that up! I didn't realize until I started the job that the pay was only US$24,000 per year. Granted, this was in 2000, but I was living in expensive New York, and my rent was $900 per month. Do the math! In fact, we will explore debt-to income ratio in the next chapter! Additionally, this job was over an hour away in New York City, and it required a daily commute by train, which was about $200 per month. I attended the first two days of work, but when I realized all too late that I would have great difficulty even making ends meet financially, I just started skipping work again and sleeping all day. I found sleeping was just a way to avoid dealing with my life. Within a week, I received a letter stating that I had been dismissed from that job.

Let me summarize my first year after college. I managed to lose three jobs, fell flat broke financially, almost got evicted from my apartment, and allowed myself to drop into the abyss of a terrible relationship. I would say failing a math class pales in comparison to my composite failure. However, I did begin to turn my life around. I found another teaching position, in an elementary school, in which I excelled. I pleaded with my landlords to allow me to stay, and I began to care for the apartment as I should have. I eventually ended that horrific relationship, and I also went back to school for my master's degree. It was a slow process, but I used my failures as lessons to help me succeed. The bottom line is that in life we are going to fail. However, successful people learn from their failures. Unsuccessful people do not.

Learning from Failure

So, what happens if you fail or need to withdraw from your math class? How can you learn from failure and turn failure into future success. You need to evaluate what went wrong.

Did You Lack Prerequisite Skills?

Throughout this book, I have stressed the importance of prerequisite math skills. Again, the lack of prerequisite skills is the biggest ingredient for failure in math. A student's work habits or even intelligence are moot if he or she does not have the proper prerequisite skills. Review the topics in the class that you covered and ask yourself some questions:

- Did you struggle with basic information that the professor expected you to already know or that the professor quickly reviewed?
- Did you feel lost as soon as a new topic was introduced?
- Were there topics that you could almost understand, but there was prior knowledge that you needed that kept you from completing the problem?

If so, Why did this Happen?

How could you be placed into a math class where you are lacking prerequisite information? The placement test is an imperfect instrument. It may be possible for students to guess correctly. Oftentimes, placement exams are not designed by the individual college; they are generic exams that several colleges utilize and may not accurately reflect the specific math content in your college.

As discussed in Chapter 2, many colleges now utilize multiple measures to place students into college classes. For example, your high school grade point average (GPA) may have allowed you to place into a higher math class. While a higher GPA may reflect a good and hard-working student, a composite GPA does not address prerequisite skills for a specific math class.

What to do?

Contact your professor and ask for his or her opinion regarding your prerequisite skills for that specific class and ask for a recommendation. It is possible you may need to take a lower math class to brush up basic skills. This will add more time to your academic plan, but it may be necessary. However, an additional class may not be necessary. You may be able to work, with a tutor, on your prerequisite skills so that you are better prepared for the class. Design a list of the topics that gave you difficulty and work with your professor or a tutor to focus on the prerequisite skills you need in those topics to be successful.

Did You Devote Enough Time to Your Studies?

The hours of the day can fill up quickly. Between work, family, and other obligations, it is easy to lose time for studying and completing homework assignments.

> When I would get home, someone always needed something, whether it was my husband or my kids. I just did not have time for my schoolwork.
>
> Joyce

> I just didn't work on my online assignments the way I should. I was supposed to finish about two assignments each week, and I just slacked off. It got time to take the test, and I would try to cram as much as possible.
>
> Ana

> My professor told us the online assignments we needed to complete [outside of class]. I started working on them at first, but then I just stopped. I just got overwhelmed with my other classes and everything else.
>
> Peter

> I would struggle in class, but I didn't spend the time I needed to [outside of class] to get the help I needed or work on stuff.
>
> Alan

Please revisit my suggestion, in Chapter 3, for mapping out a weekly schedule, which includes protected time for your studies. However, take a closer look at what entities took you away from spending time on your math. Was it other coursework? Was it your job? Was it family obligations? Did you just not feel like working on math?

You may need to make some lifestyle changes. Talk with your family members, for example, about your need for more time on your studies. If it was lack of motivation, repeat to yourself your desire to complete a college degree and why you are doing it.

Were You in the Wrong Modality?

There are a variety of modality offerings (in-person, online, virtual, etc.). This serves a diverse student population, but students may find themselves in the wrong modality and not realize this until it is too late.

> *I thought an online class would be easier. I figured I would save time by not having to drive to class and sit in class. Wrong! I didn't realize how hard it would be to learn the math on my own with videos and reading the [text]book. I fell so far behind and fast! I needed to be in a regular [in-person] class.*
>
> Tiffany

> *I took a math class where it was online, but we met online as a class [a virtual class]. Even though my professor would lecture, I felt like I was on YouTube. It was hard to concentrate. And I'll admit, I would have my camera off, and I would just get distracted and do other things. I needed to be in a classroom [in-person class] to keep me focused.*
>
> Ariel

> *I started in a face-to-face [in-person] class, but I just had to miss so many classes because of my family and sometimes my job. My son gets sick a lot, so I had to miss class. I liked my instructor, and I understood what was going on in class, but I missed too many exams and homework assignment deadlines. Online just worked better for me. It was a lot of work, but I could set my own hours. I could be home with my son and watch videos for my class, or I could stay up late and do my work.*
>
> Brenda

> *I could not learn math in a lab [emporium model]. When we got into hard stuff, like harder algebra, I needed someone to explain it to me. I just couldn't learn it on my own.*
>
> Andrea

If you believe that you were unsuccessful in your math class because of the modality, here are some questions for you:

- What was it about the modality (in-person, online, emporium model, etc.) that hindered your success?
- What modality do you believe would be a better fit?
- More specifically, why would another modality work? Why do you feel you could master the same content in a different setting?

Do You Need to Change Your Pathway?

There is also the possibility that you may need to alter your math pathway, and this may impact your career pathway. College algebra is a difficult course for many students. The difficulty stems from college algebra being a fast-paced course that is full of multiple prerequisites. More specifically, to master the content in college algebra, students must be fluent in prior content.

I tried, but I just couldn't get through college algebra. The class just went so fast, and I couldn't keep up. It was so frustrating, because I didn't understand enough of what I needed to know [prerequisites] to learn the new stuff.

Emily

The class [college algebra] was just too hard for me. I tried, but it moved too fast, and I just didn't understand it. I tried it a couple of time, and I just kept falling behind.

Jerry

I was a business major, and I had to take the Finite Math. The problem was all the algebra in that class. I tried the class three times, and I always had to drop it. The really bad part was I had to take that class just to take business calculus, and there was no way I was going to pass that. I thought I was just stupid.

Tiffany

This is going to sound stupid, but I'm a theater major, and I tried college algebra for my math requirement, because I thought "algebra" sounded easy. I mean, I had taken algebra before. I thought it was like x + 10 = 20. What is x? Guess what? It's not easy at all.

Nancy

So, what was the solution for these students?

I was taking college algebra because I wanted to become a high school math teacher. I really wanted to help kids who had trouble in math, just like me. But my advisor made me realize that I could still help kids, just younger kids, so I changed my major to elementary education, and then I only needed to take the math for elementary school teacher's class [teacher preparatory math] for my degree.

Emily

I wanted to go into forensic science, so that's why I was taking college algebra, but I realized I could do just as much good in criminal justice, and that [criminal justice] only required statistics. Statistics just made so much more sense to me. Maybe it's because you could apply it to real-life situations or maybe because we could just start from the beginning, and it didn't always feel like I was behind, but it worked.

Jerry

I had to do a lot of thinking. I struggled in the [Finite Math] class because I needed a lot of other algebra, and they just put me in Finite Math with the booster class, so I never got the chance to learn all the stuff I needed. I still wanted to go into the business field. But I never knew I could choose a different path in business that only required statistics. I was originally an economics major, but then I talked to some people in the business field, and I realized I really wanted to go into marketing and advertising, which at my school only required a couple of statistics classes. Statistics just worked so much better for me. It was hard, but I could start from the beginning and work my way up.

Tiffany

I wanted to stay away from Quantitative Reasoning. It sounded like it was so hard, but my advisor recommended that class, because it was for people who don't need a whole lot of math classes like me. I took the class, and it made so much sense. First off, it wasn't like college algebra, where I was behind from the first day, and I loved all the real-life stuff.

Nancy

Something to Keep in Mind Regarding Pathways

Hopefully, these examples helped if you need to alter your academic pathway. Keep in mind, however, that pathways can get complex. College algebra should be reserved for students who are in the STEM (Science, Technology, Engineering, and Mathematics) pathway. Quantitative reasoning and introduction to statistics should be required for non-STEM students. Nevertheless, there are times when program directors may still require college algebra for non-STEM students. This can happen for a variety of reasons. Oftentimes, the degree requirements are outdated. In other cases, such program directors may not understand what quantitative reasoning is. However, I hope we are gravitating toward a time where only STEM students will be required to take college algebra.

Did You go at it Alone?

Earlier in the book, we discussed the benefit of working with peers. Many students benefit from studying together for exams or even on a regular basis to complete homework assignments or simply keep up with the material. When you are working with peers and helping each other, you will likely find yourself explaining various math content to others, and this will help you understand the material better. Again, in an online class, try to arrange study groups in the forum. In an in-person class, if you are reluctant to ask others to join a study group, ask your professor to help you arrange study groups. Students can sign up if interested. Take the lead!

Generic Questions

Hopefully, the questions above provide you some food for thought regarding your difficulties in your math class. However, here are some generic questions you need to reflect upon after an unsuccessful attempt in a math class:

- Was I prepared for this class from Day 1? More specifically, was I aware of the prerequisites for this class, and did I ensure that I had at least some of them mastered?
- Did I devote enough time each week to reviewing my notes and completing my homework assignments?
- Did I ask for help immediately if I did not understand a concept?
- If I could go back in time, what could I change to help me pass this class?
- Do I feel prepared to try this class again? If not, what are the barriers to my success?

Again, failure should never be a goal, but oftentimes, to be successful, we must learn from failure. Moreover, for successful people, failure is often a small bump on a successful pathway.

Why in the World? Part 8

Why in the World do Logarithms Exist?

Logarithms did not make the list of the seven deadly topics in Chapters 5 and 6, but believe me, regarding being hated, they are up there!

Why logarithms make no sense: Grab a scientific calculator, and search for the "log" button, which stands for logarithm. Now, hit "log" and type in "272," so basically, log (272). The calculator displays 2.434568.... . What does that even mean? Doesn't it seem like some tiny person in the calculator just spit out some random number?

Let us try to understand: A logarithm is a way to express an exponent. We have discussed exponential equations earlier. For example, in $10^3 = 1,000$, 10 is the base, 3 is the exponent (or power), and 1,000 is the answer (or argument). However, we can convert this to a logarithmic equation like this: $log_{10}(1,000) = 3$. More specifically, the base of the exponent becomes the base of the log, and the exponent and the argument switch places. Therefore, a logarithmic equation is equal to an exponent. We can do this with any exponential equation: How about? $7^2 = 49$. For a logarithmic equation, this converts to $log_7(49) = 2$.

We have not, however, explained how log (272) magically gives us 2.434568... Let us start with something more basic. How about log (100); that is 2. How about log (10); that is 1; How about log (1); that is 0. What is happening here? For log (100), this is a logarithmic equation. When we do not see a base below the log, we assume it is base 10. Therefore, this equation is $log_{10}(100) = x$. If we convert this to an exponential equation, it becomes $10^x = 100$. We know that $10^2 = 100$, so $x = 2$. When we typed in log (100), we were asking the calculator, "10 to what power is 100?" That is why the calculator responded with "2." Circling back to log (272), let us understand this as a logarithmic equation, which is $log_{10}(272) = x$.

Converting this into an exponential equation, we get $10^x = 272$. Therefore, we are asking the calculator, "10 to what power is 272?" Again, the calculator gave us 2.434568... That means $10^{2.434568} = 272$ or approximately 272. If we check it on the calculator, we see it is true. When we type in log of any number into the calculator, we are asking the calculator "10 to what power is that number?"

Hopefully, this explains more about how logarithms work. However, we have not answered why logarithms exist. There are many real-life examples, but let us focus on earthquakes, which are typically measured on the Richter Scale. More specifically, the Richter Scale measures the size or magnitude of an earthquake. While watching the news or reading information online, you may have heard or read that an earthquake measured 5, 7, or 8 on the Richter Scale, but what does that mean? Earthquakes can be horrific, and 5, 7, or 8 are not exactly gargantuan numbers. Suppose that three states experienced earthquakes: Alaska, California, and Hawaii. Here are the sizes for each on the Richter Scale: Alaska—5, California—7, and Hawaii—8. These numbers are not 5, 7, and 8; they are 10^5, 10^7, and 10^8. More specifically, $10^5 = 100,000$, $10^7 = 10,000,000$, and $10^8 = 100,000,000$. On the surface, the numbers 5 and 7 seem close. However, an earthquake that measures "5" is considered moderate, with some shaking and minor damage. However, an earthquake that measures "7" will cause serious damage over a widespread area. That is because the earthquake in California was 100 times greater than the earthquake in Alaska. The earthquake in Hawaii that measured "8" is considered a "great earthquake" and is 10 times worse than the earthquake in California. This is a major purpose of logarithms. Logarithms allow us to compare very large numbers in a manageable way.

Activity

- Review my story of my experience with failure. From an outside perspective, what were my key mistakes? Is there a way I could have prepared better to avoid these mistakes. What lessons could a person learn from these mistakes?
- Does the prospect of failure frighten you? Why? Is this based on past experiences with failure? If so, how did failing make you feel, and what can you learn from that failure?
- Picture yourself succeeding in a math class. How will that make you feel? What can you see yourself doing to achieve that goal?

9

When Will Math Help Me in My Life?

Ask any math educator the question they here most from students, and I guarantee several will reply, "When am I ever going to use this [math] in real life?" Some educators may even hang their heads or roll their eyes as they recite this question. The truth is that while students often ask this question in frustration, it is a legitimate question, and we will address that in this chapter. Specifically, we will address how math can be useful in STEM (Science, Technology, Engineering, and, of course, Mathematics) fields but also for non-STEM students.

Non-STEM Students

In this section, we will address how math can be useful for non-STEM students. In other words, let us say you only need one college-level math class for your degree. How can math help you in your life? Additionally, keep in mind that these examples can benefit both non-STEM and STEM majors.

Helping Others

As I have mentioned numerous times in this book, math is a struggle for many people in various areas of education. Mastering some math skills can put you in a position to help others.

> After I passed my classes I became a math tutor at my college. I loved it. Because I had struggled so much in math, I feel like I was able to relate to students. And I could tell that the students knew I really understood them. I wasn't just someone else teaching them math. I wasn't a math major, but I was able to help students

 DOI: 10.1201/9781003614142-9

in their algebra classes, and that was pretty cool, and I made some money while going to school!

Kathryn

My nephew was in 6th grade and was really having trouble with some basic algebra, especially the signed numbers. My sister and brother-in-law were pulling their hair out, because they couldn't help him. It was a screaming match in their house every night. I offered to help my nephew after I passed algebra, and things got so much better. He said I explained it better than his teacher.

George

At the church I go to, they started an after-school program where kids do their homework. I started to volunteer, and so many of the kids, like in 4th or 5th grade were having problems with fractions. I was like, "Wow, I can do this now and actually help other people." I think I was able to help the kids better because I had so much difficulty with math.

Alicia

For this next testimonial, if you are single and dating, I cannot promise this will happen to you, but it is a nice story.

I was on this dating app and going on a lot of blind dates that really went nowhere. Anyway, I went on this one date with this really beautiful girl, and I was really into her, but I felt like the date wasn't going anywhere. I was just having trouble coming up with things to talk about. Then, all the sudden I mentioned how I was in college and had trouble with math but finally passed my math classes. She perked up and mentioned how she had failed her developmental algebra class three times. That really broke the ice, and then she asked if I would tutor her! Of course I did. I even found myself brushing up on math I had done, just so, you know, I could tutor her well. Long story short, we have been dating for a year now.

Derek

Managing Money

Living on a budget and avoiding falling into debt can be a challenge. However, mathematics can help with this. Understanding the debt-to-income (DTI) ratio can be salient to managing money. There are two types of DTI: front-end DTI and back-end DTI. Front-end DTI refers to monthly housing expenses. More specifically, for example, let us say, Alan's gross monthly income is \$4,000, but he wishes to rent an apartment for \$1,400, the front-end DTI would be 35%. This is calculated by: $\dfrac{montly\ housing\ expense\,(1400)}{gross\ monthly\ income\,(4000)} = .35$, which converts to 35%. Is this good or bad? Typically, banks, leasing agents, or landlords, require a person's front-end DTI to be at most 28%. Therefore, Alan would likely not receive the apartment. Even if Alan did receive the apartment, more than one-third of his income would be consumed by his rent. This may not seem like much, but keep in mind that gross pay is pre-tax, which means that government and state taxes plus other deductions have not been applied.

Back-end DTI refers to all monthly expenses compared to a person's gross monthly income. Let us consider another example. Samantha earns $4,650 per month. She rents an apartment for $1,100. Samantha purchased a car and has monthly payments of $390. She also has a monthly credit card bill of $320 and monthly student loan payments of $290. What is Samantha's back-end DTI? This is computed by: $\dfrac{all\,monthly\,expenses : 1,100 + 390 + 320 + 290}{4,650}$. This simplifies to .4516, which converts to 45.2%. Is this good or bad? Typically, a person's back-end DTI should not exceed 36%. Therefore, with a high back-end DTI, Samantha could find herself in debt. In summation, when people are not cognizant of their DTI, they can struggle in their finances.

Let us approach DTI another way. Dan earns $70,000 each year, which is a gross monthly income of $5,833.33. However, he has the following monthly expenses: $1,400 in rent, $350 in credit card payments, and $250 in student loan payments. Dan wants to apply for financing to purchase a new car. The company agent will only approve his loan if his back-end DTI is no higher than 36%. Considering what Dan earns and his other monthly expenses, how much per month can Dan pay for his car? Let us write down what we know:

So, x = what Dan can pay per month for his new car.

Can we translate this to an equation, considering what we know about back-end DTI?

$$\frac{1,400 + 350 + 250 + x}{5,833.33} = .36$$

Again, back-end DTI is the ratio of all monthly expenses compared to monthly earnings. The numerator lists all of Dan's monthly expenses, along with "x," our unknown which is his potential car payment. Why did we set the equation equal to .36? Because we are asking: What value (of x) would make this equal to 36%? More specifically, what car payment would make this back-end DTI 36%?

How do we solve for x? We encountered a similar equation in Chapter 6. Let us simplify the numerator:

$$\frac{2,000 + x}{5,833.33} = .36$$

We can even think of this as:

$$\frac{2,000+x}{5,833.33} = \frac{.36}{1}$$

This is a proportion where we can cross multiply to remove the fractions. We can multiply $(2,000+x)(1)$ and $(5,833.33)(.36)$. This gives us: $2,000+x = 2,100$. Please note that I rounded \$2,099.99 to \$2,100. Now, subtract 2,000 from both sides to isolate x.

$$2,000+x = 2,100$$

$$-2,000 = -2,000$$

$$x = 100$$

This means Dan will likely only be financed for \$100 each month for a new car. Unless Dan puts a lot of money down on the new car, \$100 per month is not much.

Think about this. While the cost of living varies by state, Dan's salary of \$70,000 is above the national median salary. On the surface, you would think Dan would be able to pay more per month for a new car. However, his back-end DTI, which includes his monthly expenses provides the reality of his financial situation and the additional debt he can accrue. In fact, Dan may have assumed that because he earned a good salary that he could have afforded a higher car loan. This is why an understanding of DTI is so imperative.

Predicting the Cost of Living

Consider the following situation. Denise earns \$50,000 a year. Five years later, she earns \$55,000 per year; yet she is struggling in her finances. Why? After all, she is earning more money, but is she keeping up with the cost of living? Inflation is generally a given. Expenses such as groceries and gas prices will likely increase. However, if inflation increases at a greater rate than a person's income, that becomes problematic.

Here is a formula that is used to calculate the cost of inflation: $FV = PV(1+r)^t$. Let us identify the variables: FV = Future Value, PV= Present Value, r = rate of inflation over a certain amount of time, t = the time being considered. More specifically, for the above-situation, 50,000 would be the PV. Let us say the average rate of inflation for the five-year period was 4.5%; therefore $r = 0.045$, and since this is a five-year period, $t = 5$. Now, let us determine how much Denise's salary would be worth five years later: $FV = 50,000(1+0.045)^5$, which leads to a Future Value of \$62,309.10. So, while Denise earned \$5,000 more in those five years, it was not enough to keep up with the rate of inflation.

Being a Conscientious Consumer

Saving money and cutting costs are good ingredients to a healthy financial situation. While cost-cutting when shopping may seem small and even insignificant, continuing to be a conscientious consumer adds up over time. Consider the following example:

In a retail supermarket, 24 bottles of water cost $5.99. In a wholesale store, 40 bottles of water cost $6.99. Which is a better deal?

Certainly, it seems like the consumer is paying more at the wholesale store; however, we want to determine the cost per bottle. We do this by dividing the overall cost by the number of bottles. At the retail supermarket ($\frac{5.99}{24}$), the consumer is paying 25 cents per bottle. At the wholesale store ($\frac{6.99}{40}$), the consumer is paying 17 cents per bottle. Therefore, the consumer is getting the better buy at the wholesale store.

Let us consider another similar example.

In a retail supermarket, Tide costs $16.99, and this will allow for 64 loads of laundry. In a wholesale store, Tide costs $19.99, and this will allow for 100 loads. Which is the better buy? We figure this out by computing cost per load. At the retail store $\left(\frac{16.99}{64}\right)$, the consumer is paying 27 cents per load. At the wholesale store $\left(\frac{19.99}{100}\right)$, the consumer is paying 20 cents per load. Again, the consumer is getting the better buy at the wholesale store.

Computing Depreciation Costs

In life, when you make purchases, those items either appreciate or depreciate. Houses, certain artwork, fine wine, or classic coins, as some examples, appreciate over time. However, other items such as cars, motor homes, and electronics, such as TVs and video game consoles, depreciate. If you want to sell an item you purchased, you will want to have an idea as to how much you will receive for it. Consider the following situation. You purchase a new Toyota Corolla for $22,865. You have now owned the Corolla for three years. Table 9.1 shows how much the car is worth after each year.

In Chapter 5, we discussed slope and the equation of a line and how we could use the equation of a line to predict future values. That is what we will do here! More specifically, the equation of a line is $y = mx + b$ where "m" is the slope and "b" is the y-intercept. However, recall that the slope of a line is a constant rate of change. Look at how the car is depreciating—it is not by the same amount each year. So, how can we construct an equation

TABLE 9.1

Car Depreciation

Year	Car value
0	$22,865
1	$20,700
2	$18,865
3	$17,000

of a line, and more specifically identify a slope, if there is not a constant rate of change?

Let me ask you this. Do you think in real life there is always a constant rate of change? No. In this case, the value of a car is not going to decrease by the exact same amount each year. Therefore, we will construct an equation of a line, but it will be a line of best fit. In other words, we will come up with the best possible equation of a line. To do, this, we can use a tool such as Microsoft Excel or Alcula.com. Consequently, I was able to obtain $y = -1,943x + 22,772$ as the line of best fit. Think of it as the best possible slope given the circumstances. Also, notice the slope is negative, as the value is decreasing.

Now, let us use this line to help make decisions for the future. What if you wanted to sell the car after five years? How much could you get? Well, "x" represents the years, and "y" represents the value of the car. So, $y = -1,943(5) + 22,772$. This gives us 13,057. Therefore, after five years, we can predict the car will be worth $13,057. How about if you wanted to ensure that you receive $7,000 for the car? After how many years should you sell it? In that case: $7,000 = -1,943x + 22,772$, Let us get x by itself:

$$7,000 = -1,943x + 22,772$$

$$-22,772 \qquad\qquad -22,772$$

$$-15,772 = -1,943x$$

Now divide:

$$\frac{-15,772}{-1,943} = \frac{-1,943}{-1,943}x$$

$$8.1 = x$$

So, just after eight years, we can predict that the car will be worth $7,000.

Statistics can do a Lot of Explaining

Here are the results from the first exam in a new math class that students are taking their junior year of high school: 24, 35, 40, 45, 66, 68, 75, 78, 83, 85, 88, 90, 92, 94, 95, 98, 99, 100, 100. Furthermore, a new teacher is instructing this class. I want to focus on the student, Johnny, who received a 40%. Johnny's mother cannot believe he received a 40% and is complaining to the principal about the teacher. Certainly, Johnny scored poorly, but who is to blame? A box and whisker plot, which is a part of statistics, can not only shed some light on this, but provide a solid argument for the winning party. To construct a box and whisker plot, we need to determine five values: the lowest score, the lower median (Quartile 1), the median, the higher median (Quartile 3), and the highest value. The lowest (24) and highest (100) values are the easiest to determine. To determine the median, we choose the middle number, which is 85. To obtain the lower median, we look at the numbers to the left of the median, which are 24, 35, 40, 45, 66, 68, 75, 78, 83. The lower median is the median of those numbers, which is 66. To obtain the higher median, we look at the numbers to the right of the median, which are 88, 90, 92, 94, 95, 98, 99, 100, 100. The higher median is 95. To summarize, here are the five important values: Lowest—24, lower median—66, median—85, higher median—95, highest value—100. From this we can construct a box and whisker plot (see Figure 9.1).

As you can see, the whiskers on the left and right represent the lowest (24) and highest (100) values, respectively. The red line inside the box represents the median (85), Finally, the left side of the box starts at the lower median (66) and extends to the higher median (95).

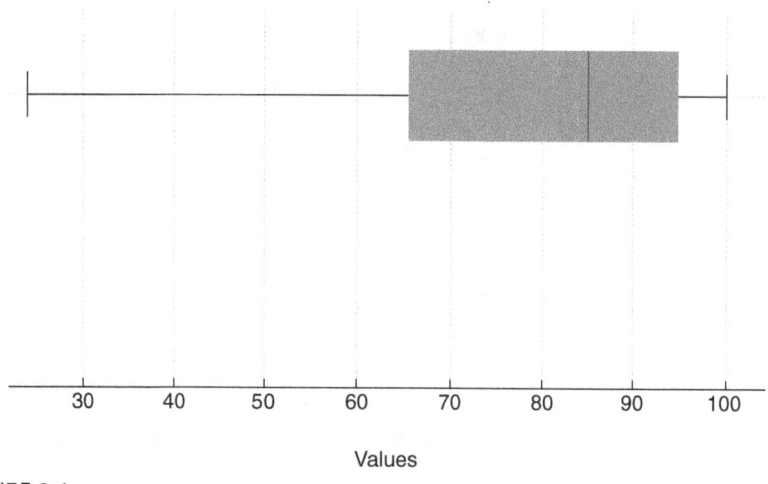

FIGURE 9.1
Box plot

What does all this tell us about the conflict presented earlier? Here is what box and whisker plots tell us. The lowest 25% of the values are below the lower median (66). The middle 50% of the values fall between the lower median (66) and the higher median (95). The highest 25% of the values are above the higher median (95). Where did Johnny score? Johnny's grade of 40% was in the lower 25% of the class. Consequently, half of the class scored between 66 and 95. Furthermore, 25% of the class scored above 95%. I would say the new teacher is doing a fine job, and Johnny has some studying to do and some explaining to mom! The bottom line is a box plot can settle a dispute with hard evidence.

Higher-Level Math Classes

Later, we will examine how studying mathematics can help STEM students. We will look at a few STEM pathways and the mathematics required for each pathway. Before that, however, let us examine the descriptions for some higher-level math classes we will be referencing, as STEM students must often complete such classes.

Trigonometry

Depending on the college, trigonometry may be a part of college algebra, pre-calculus, or it may be a stand-alone class. Trigonometry is part of geometry and focuses on angles. To be successful in trigonometry, students need a basic understanding of angles, more specifically, an understanding of the laws of congruent angles. Students also need a mastery of algebraic concepts such as linear equations, quadratic equations, and factoring.

Pre-Calculus

The purpose of pre-calculus is to prepare students for the rigors of calculus. There may be some overlap between college algebra and pre-calculus. Students focus on various types of functions (polynomial, exponential, and logarithmic). Other topics may include conic sections, which are the studies of circles, ellipses, parabolas, and hyperbolas. Some colleges also include trigonometry as part of pre-calculus. To be successful in pre-calculus, students must have a deep understanding and mastery of all algebraic concepts.

The Calculus Sequence

Most colleges offer calculus in three semesters (Calculus 1, Calculus 2, and Calculus 3). Not all STEM students need to complete all three semesters of calculus. We will discuss how to "survive calculus" further into the chapter; however, calculus, in simple terms, focuses on the rates of change and growth in everyday life. Calculus is filled with complex topics such as limits, derivatives, integrals, and vectors. Students must have a deep mastery of algebra and trigonometry to be successful in calculus.

Differential Equations

Differential equations is an advanced undergraduate math class that typically follows the calculus sequence. In fact, I have heard several math faculty refer to this class as "Calculus 4." Students must master all concepts in calculus to be successful in differential equations. Most non-math majors do not need to complete differential equations; however, some STEM degrees do require this class.

Linear Algebra

Linear algebra focuses on matrices, transformations, and vectors. Yes, that sounds complicated, and it is. Many colleges require students to complete two semesters of calculus before attempting linear algebra. However, the content in undergraduate linear algebra does not rely much on calculus. Oftentimes, the topical prerequisites for linear algebra exist in a finite mathematics class where students study various types of matrices operations. However, as linear algebra is a difficult class, students need a high level of mathematical maturity. Therefore, assigning two semesters of calculus as a prerequisite assures such maturity.

Proof Writing Classes

After completing calculus and differential equations, mathematics classes become more abstract. In such classes, students must complete rigorous proofs. In completing a rigorous proof, students must complete an argument using various facts. For example, students must prove the Pythagorean theorem. Therefore, these proof-writing classes guide students in writing a rigorous proof. To be successful in proof writing, students need a deep understanding of algebra, geometry, and calculus, as students must use facts and theorems from these classes for rigorous proofs. Introduction to analysis is often one of the first undergraduate math classes to rely on rigorous proof writing.

STEM Students

What are some STEM career pathways? What are the benefits to jobs in these fields? Additionally, how much math is required in these fields? We will answer those questions in this next section.

High School Math Teacher

Considering that many students had negative experiences in math while in high school, some aspire to become high school teachers to make the experience better for other students.

What Does the Job Entail?

Let us hear from two high school math teachers regarding their jobs.

> *I love my job, but it's involving. I get up at 5:30 in the morning. I teach from 7:30 to 2:30 each day. I teach six classes each day, and each class is about 40 minutes. I also supervise lunch. I stay after school to help students. I grade lots of papers and homework at home. I'm always planning lessons. Sometimes it feels like I work round the clock every day, but I love what I do. I love working with my students, and I love watching them succeed.*
>
> Lucille, High School Math Teacher

> *The job can be a grind. I love working with my students, but classroom management can be a problem. [High school] students often just don't want to learn, so it can be tough to get through to them. I also work with many students who have family problems as well, which serve as extra barriers. But I know I am making a difference in my students' lives.*
>
> Harry, High School Math Teacher

How do I get to be a High School Math Teacher?

Most colleges or universities offer a degree program that is generally titled, "Mathematics Secondary Education" or something similar. Like most majors, students must complete their basic core requirements (e.g., psychology, sociology, science, English). Students must also complete a variety of higher-level math classes (discussed later) and education classes.

To teach in the public sector, future teachers must become licensed in a specific field. Each states' Board of Regents sets their own requirements for licensure. However, in addition to completing a degree, students must also pass various state exams to become licensed to teach high school mathematics. For example, in Ohio, teacher candidates must pass the Ohio Assessment for Educators (OAE) focusing on professional knowledge and the OAE focusing on mathematics.

How Much Math do I Need?

High school math teachers generally teach algebra, geometry, trigonometry, pre-calculus, and possibly calculus. Since calculus is generally the highest class offered in high school, teaching candidates should not need to master much more than calculus, right? Wrong! A bachelor's degree in mathematics secondary education is rigorous. Teaching candidates must complete all levels of calculus, differential equations, and much more abstract classes such as complex variables and introduction to analysis.

Engineer

There are various types of engineers, but ultimately, engineers analyze, design, build, and create to solve various problems and make life easier. There are civil engineers, electrical engineers, environmental engineers, and chemical engineers. Aviation, which is the study of flight, is another type of engineering.

What do Engineers do?

Let us hear from three people in the engineering field.

> *I'm an environmental engineer, and I study water quality in my city. Basically, I try to ensure that our city has the best water quality possible. I deal with problems that arise and look for solutions. I've always cared about the environment, so I love knowing that I am working to help people and giving people a better quality of life.*
>
> Clair

> *I'm an electrical engineer, and I work for Tesla. I'm so excited about electric cars, because I care about the environment. People don't realize how bad carbon monoxide is for our air. In a nutshell, I work to improve Tesla cars and make them safe and more convenient to use for people.*
>
> Lenny

> *I always wanted to be a pilot, and I love to fly. It's my first year as a pilot, and I'm so psyched. My degree in aviation was so important. I learned so much and got so much hands on experience!*
>
> Dylan

It is noteworthy that the engineering field is growing and is in high demand. In fact, as industries, companies, and general needs continue to grow, there is an average need of 400,000 new engineers each year. In fact, the need for new engineers will increase by 13% from 2023 to 2031. Why is there a shortage of engineers? As we will discuss below, the academic degree requirements are rigorous, but as always, with the proper prerequisite skills and hard work, students can complete such degrees.

How Much Math do I Need to be an Engineer?

This is a complex question. In general, most engineering fields require students to complete all semesters of calculus. Electrical engineers, for example, may need higher-level math classes such as differential equations. Additionally, engineering students likely need to complete math-engineering classes. Other engineering fields require classes in probability and statistics. Furthermore, students must also complete advanced science classes, which utilize applied mathematics, in chemistry and physics in fields such as environmental or chemical engineering. More specifically, these are classes where math is applied to specific engineering fields. Not all engineering fields require calculus, however. Many aviation programs only require trigonometry as the highest math requirement.

Let us address a common question. With the existence of advanced technology and the rapid growth of artificial intelligence, why do engineers need to master math?

> *Here's the thing. We have computers and machines to compute the math, but I still need to problem solve, and so much of what I do still involves concepts like limits, derivatives, and integrals, so I need to be able to understand and interpret the information. So, no, a person who has not studied advanced math could not do my job. That's a scary thought, actually.*
>
> Lenny

Computer Scientist

A computer scientist works in various industries; however, computer scientists use their skillset in programming, system design, data analysis, and communication to problem solve and improve technology and the quality of life for humans.

> *I love my job. I test several new programs for my company to see which ones work the best and are most efficient, and I also do a lot of problem solving, and I do my best to get programs to run efficiently. It's a high-pressure job, because there is a lot at stake. If our programs do not run well, our company doesn't run as well, and we can't serve our clients well.*
>
> Nick, Computer Scientist

> *I work on various anti-virus programs to help people keep their laptops and phones safe. My job involves using a lot of algorithms and coding. I love my job, because in this day and age, people rely 100% on technology. I know I am making people's lives easier and less stressful.*
>
> Mary, Computer Scientist

Keep in mind that while being a computer scientist holds a great deal of responsibility, it can also be a lot of fun. For example, video gaming is fun; isn't it? Who designs such games? People who are creative and have expertise in programming and coding. Yes, computer scientists can design video games.

How Much Math do Computer Scientists Need?

In general, students must complete at least two semesters of calculus and linear algebra. Linear algebra prepares students for the intense programming skills they will need for computer science. Linear algebra is also imperative for machine learning. Additionally, computer science majors need to complete math classes such as discrete mathematics and advanced statistics. Ultimately, these math classes offer abstract reasoning and the development of logic, which is imperative in the field of computer science.

Astronomy

Astronomers study the planets, stars, as well as origins of the universe. Astronomers play a huge role in our understanding of the universe. Astronomers design new instruments and maintain equipment. They also develop software to interpret images captured by satellites.

Astronomy sounds complicated, but think about satellites. Satellites help meteorologists predict weather and important events such as earthquakes, hurricanes, and tsunamis. This can protect and save lives. How about this? Because of astronomy, the United States became the first country to put a man on the moon. In fact, as of 2024, the United States have sent over 300 astronauts into Outer Space. The field of astronomy is also in high demand. In fact, the marketplace for astronomy is expected to grow by about 8% between 2022 and 2032.

What Kind of Math do I Need to be an Astronomer?

To complete a degree in astronomy students must complete all semesters of calculus. Additionally, students must complete linear algebra as well as a proof-writing class. This is in addition to completing several high-level science classes with applied mathematics.

Can I Survive Calculus?

A common theme to completing degrees in STEM is that students must complete calculus and, in many cases, multiple semesters of calculus. For students who have struggled with math but wish to pursue a career in the STEM field, completing even the first semester of calculus can seem daunting. A student may think, "I've struggled with basic algebra or even arithmetic; how in the world can I pass calculus?" This is an understandable question, and we will unpack that in this section.

There is no question calculus is a rigorous class and sequence. Earlier in this chapter, we discussed the concepts in calculus as well as the prerequisites. So, how do students survive in calculus? Let us hear from them:

> *Calculus was hard, but I got through it. It feels like a different language with different symbols and letters. It felt like I was learning math all over again.*
>
> Jose

> *You want to pass calculus? You need to know your algebra and your trig[onometry]. That was where our class struggled the most. It used to drive our [calculus] professor crazy. He was like "You people need to know your algebra to pass this class."*
>
> Victor

> *Calculus 1 [first semester] took me two tries. I learned my lesson the first time. First off, the class went so fast, faster than pre-calculus ever did. So, I fell behind so fast. I didn't spend as much time studying as I should have. And I didn't know my algebra and really my trig as much as I should have. I took trig as part of pre-calculus, so I kind of just learned it enough to pass the class, but I did not know as much trig as I should have. So, after I failed calculus, I spent some time with a tutor going over the algebra and trig I needed.*
>
> Robin

> *My biggest mistake was I had a lot of trouble in Calculus 1 and basically got by with a "C," and then I assumed everything would be all right in Calculus 2, but guess what? I was failing Calculus 2 so badly, I had to drop the class. So, if you are having trouble in Calculus 1, your problems aren't just going to magically go away.*
>
> Larry

There are some common themes here, and they are all true.

- Calculus is a very fast paced class, faster than any previous math class.
- Calculus relies heavily on prerequisite concepts from algebra and trigonometry (as mentioned earlier in this chapter as well).
- Calculus, itself, is very progressive in that if students do not master basic prior calculus concepts, they will not master future concepts.
- Calculus takes on a new language of math.

Let us unpack all these concepts. Earlier in this book, we discussed that college math classes are indeed fast-paced, and it is imperative for students to develop a schedule where they allow "protected time" for their math studies and homework. This intensifies greatly in calculus. Calculus classes tend to meet four or five hours each week, and students need to spend at least 10–15 hours each week outside of class completing homework or studying their material. A single calculus class is flooded with content. Additionally, students must learn the nuances of calculus as well. Keep in mind that a student who is ready for calculus generally has a high level of math maturity. More specifically, at this point, most students know how to study and prepare for math. However, the intensity and pacing of calculus are still surprising for many students.

I have mentioned this numerous times in this book, but it bears repeating one final time. Lacking the prerequisite skills is a major ingredient to failing a math class, and that is especially true in calculus. When attempting calculus, inquire from your professor, or even others who are familiar with the calculus class, which algebra content you must have mastered. For example, calculus students should understand linear and quadratic equations, polynomial functions, and composite functions. Additionally, students must understand the laws of exponents as well as all aspects of factoring. Fractions do not go away. Students must understand how to simplify algebraic and complex fractions. I realize these concepts may not make sense until you have studied them, but hopefully this demonstrates that students need a deep understanding of algebra. Furthermore, students must have a thorough understanding of trigonometric functions as well as similar and congruent angles.

Calculus can feel like a new language for students. Prior math classes tend to be more concrete and involve top-down problem solving. More specifically, a student is given an application, and he or she employs various steps to arrive at an answer. This is still the case with many calculus applications; however, in calculus students are also presented with proofs and theorems. More specifically, students are presented with arguments that lead to various mathematical theorems. There was a time when many calculus professors would require students to recite proofs on an exam; however, in the present day most do not. Students are simply introduced to such proofs when learning a new concept. However, the content in calculus simply has more of an abstract feeling when compared to previous math. When starting calculus, students are usually first introduced to limits. Consider Figure 9.2, which focuses on an introductory problem in calculus.

This problem looks daunting but let us first dial it back to algebra. Would you agree that where the x-value is 2, the y-value is 4? More specifically, if you take a pen and trace it to 2 on the x-axis, but if we move upward, we meet at 4 on the y-axis. However, this question is asking: What is the limit of the function as x approaches 2? What does this mean? The graph that is

Using the graph below, find the $\lim\limits_{x \to 2} x^2$

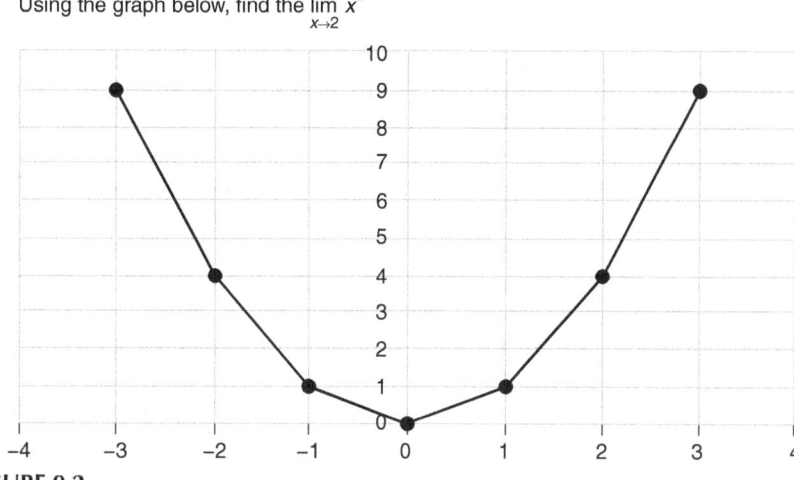

FIGURE 9.2
Limits

in Figure 9.2 is a function, which is a specific type of a relation (points on a graph). The name of the function (all functions have a specific name) is $y = x^2$, just like all lines have names (as discussed in Chapter 5 and earlier in this chapter). The question is asking, as we move toward 2 on the x-axis, what y-value does the function approach. More specifically, if we take a pen and move toward 2 on the x-axis, from either direction, the function is moving toward 4 on the y-axis. This is part of the language of calculus. We are not interested in the exact meeting point of an x and a y coordinate, we are asking as we move toward a value on the x-axis, what value are we moving toward on the y-axis.

Two Types of Students

Let us consider two ways of entering calculus. First, there is Group A, the students who complete four years of college preparatory or even honors math in high school. This likely includes two years of algebra, one year of geometry, and a year of pre-calculus. Then, there is Group B, the students who did not see much success in high school math or basically entered college enrolling in developmental math, and these students needed to complete college algebra and trigonometry before entering calculus. Which group has the advantage?

I want to be clear that with hard work and the mastering of prerequisites that any student can be successful in calculus. However, Group A has the advantage for success in calculus. Throughout those four years of high school, students gain a deep understanding of algebra, geometry, and trigonometry.

Additionally, they get repetition and practice over those years to deepen their knowledge base. More specifically, in that year of geometry students are introduced to rigorous geometric proofs and gain experience making a mathematical argument. The student from Group B who relies on developmental math, college algebra, trigonometry, and even pre-calculus in college is introduced to the necessary topics but is not exposed to the same kind of mathematical depth as the student from Group A. That is because college courses are typically 15 or 16 weeks. That is why these students may struggle with more obscure algebra concepts in calculus. This is especially the case for trigonometry. In high school, students gain a thorough understanding of trigonometry that may last several months. Many college courses that include trigonometry as part of college algebra or pre-calculus set trigonometry up as somewhat of a crash course. Students may spend roughly three to four weeks on the necessary trigonometry concepts but do not get the deep understanding of the topic.

What can Group B do to prepare for calculus? These students must be cognizant that they may not have as deep of an understanding of the prerequisite skills. They must use the time before calculus to practice the algebra and trigonometry that they will need for calculus.

Different Types of Calculus

As referenced earlier, not all students who enroll in calculus have the same goals.

Consequently, some colleges have started offering different types of calculus for students based on their pathways. Students who are math, engineering, or computer science majors need to complete the traditional calculus classes. This is especially true if these students need to complete differential equations or a proof-writing class. However, colleges may offer business or science majors a different calculus class.

Students who are completing business degrees that require calculus may complete a business calculus class. This class includes the major concepts of calculus; however, the class excludes trigonometry and does not include as many of the proofs of a traditional calculus class. Additionally, the class includes applications that are utilized in the business world.

The same is true for calculus for science majors, although colleges likely have varying names for this class. This calculus class still includes the major calculus concepts without as many proofs or trigonometry. Applications are contextualized toward the science fields. However, this is not as cut-and-dried as calculus for business majors. Students who are in chemistry or physics fields may still need to complete all semesters of calculus and differential equations. However, certain biology majors may only need minimal calculus, and calculus for the sciences could be a good fit.

Confidence

I want you to take a few minutes and reflect on some difficult tasks that you have conquered in your life that initially scared you? In fact, I would like you to write down these achievements and how you felt when you conquered your fears. You may not realize it, but the successful completion of each task has made you stronger and has given you more confidence. Successfully completing a math class, when you were initially afraid to attempt a math class, will give you confidence. As a person, you will become stronger. You will be able to add it to the list of personal achievements where you conquered a fear.

Let us hear from more students on successfully passing their math classes:

> *I truly never thought I could pass a college math class. I always thought I was too stupid, and other people thought I was too stupid. It was a lot of hard work, and there were some setbacks, but I did it. And no one can take that away from me.*
>
> Emily

> *I hated math, and I was afraid of math. But I did it. My advice is to make sure you understand the math you need coming into a math class [prerequisite skills]. That was the key for me.*
>
> Bob

> *Since I had always had so much trouble all my life in math, passing a college math class seemed impossible, but I did it! It's important to understand how you learn best, as a student, and use that [knowledge] to help you learn math.*
>
> Alicia

Now Go Do This!

I hope the information and stories in this book have helped. Remember, you are starting college with a clean slate, and even if you have attempted a math class and have failed, you can still begin with a clean slate. You are not programmed to be "bad at math" or "bad at numbers." Regardless of your prior academic background, race, or gender, you can succeed in math! Like any other academic discipline, math is challenging, but success is achievable. You need to arm yourself with the proper prerequisites, enough time to devote to your studies, the right attitude, and be sure to keep on top of everything.

Everyone has needed help at some point in math, even math professors and mathematicians. Everyone who has ever taken a math class has felt confused and frustrated at some point. You are not alone. So, ask for help when you need it. Do not fall behind! You deserve success and you deserve to be happy. So, go ahead and take the first step to your success. You got this!

Appendix A

Answers to Prerequisite Course Topics in Chapter 3

ANSWERS TO FIGURE 3.1

Course: Introduction to Algebra or Prealgebra
Typical Topics you Need to Understand Entering this Class:

- Addition, subtraction, multiplication, division with whole numbers.
  - 1) 1,354
  - 2) 2,010
  - 3) 14,043
  - 4) 148 R 8

- Round the following numbers to the specific place value:
  - 1) 234.6
  - 2) 768.10

- Addition, subtraction, multiplication, division with fractions.
  - 1) $\dfrac{23}{30}$ 2) $\dfrac{1}{7}$ 3) $\dfrac{5}{3}$ or $1\dfrac{2}{3}$

- Addition, subtraction, multiplication, division with decimals.
  - 1)   8.056
  - 2)   12.302
  - 3)   149.04
  - 4)   86.8

- Solving proportions.
  - 1)   1.5
  - 2)   277.5 calories

- Percentages
  1) 25.5
  2) 53.3%
  3) $169.23
  4) 4.6%

- Addition, subtraction, multiplication, and division with signed numbers.
  1) −48
  2) 15
  3) −960
  4) −12

ANSWERS TO FIGURE 3.2

Course: Elementary or Intermediate Algebra
Typical Topics you Need to Understand Entering this Class:

- Evaluating algebraic expressions.
  661

- Using the distributive property:
  −12x +36

- Simplifying algebraic expressions.
  1) $4x^3 + 3x^2 - 16x - 7$
  2) $-13x^2 + 10x + 73$

- Solving basic linear equations.
  1) $x = \dfrac{5}{3}$
  2) $x = 132$

- Solving algebraic word problems.
  1) The first number is 19. The second number is 18. The third number is 31.
  2) Jim earned $11. Bill earned $33. Tom earned $16.

ANSWERS TO FIGURE 3.3

Course: College Algebra or Pre-calculus
Typical Topics you Need to Understand Entering this Class:

- **Solving complex equations with fractions and parenthesis**
  $x = -25$
- **Factoring expressions**

  Factor the following:
  1) $(x - 9)(x - 1)$ 2) $(x - 4)(x + 4)$ 3) $4(x - 4)(x + 2)$
  4) $(x + 1)(4x - 11)$ 5) $(x + 3)(x^2 - 3x + 9)$

- **Solving quadratic equations by all three methods**

  1) Solve by factoring:
  $$3x^2 - 13x + 10 = 0$$
  $$(3x - 10)(x - 1) = 0$$
  $$x = \frac{10}{3}, x = 1$$

  2) **Solve by completing the square**
  $$x^2 - 8x - 48 = 0$$
  $$x^2 - 8x = 48$$
  $$x^2 - 8x + 16 = 48 + 16$$
  $$(x - 4)^2 = 64$$
  $$\sqrt{(x - 4)^2} = \sqrt{64}$$
  $$x - 4 = \pm 8$$
  $$x = 4 \pm 8$$
  $$x = 12, x = -4$$

  3) **Solve by using the quadratic formula:**
  $$2x^2 - 5x - 9 = 0$$
  $$a = 2, b = -5, c = -9$$
  $$x = \frac{-b \pm \sqrt{b^2 - 4ac}}{2a}$$
  $$x = \frac{-(-5) \pm \sqrt{(-5)^2 - 4(2)(-9)}}{2(2)}$$

$$x = \frac{5 \pm \sqrt{97}}{4}$$

$$x = \frac{5 + \sqrt{97}}{4} \text{ and } x = \frac{5 - \sqrt{97}}{4}$$

- **Finding the slope and equation of a line**
  1) $y = 12x + 53$
- **Solving system of equations**
  $$x = \frac{13}{7}, y = \frac{3}{7}$$
- **Simplifying rational expressions**
  $$\frac{x^2 - 3x - 2}{(x-4)(x-3)}$$
- **Multiplying and dividing rational expressions**

  1) $\dfrac{1}{x}$ 2) $(2x+1)^2$

- **Solving 1nequalities**

  1) $\left(-\infty, \dfrac{11}{2}\right)$ 2) $\left(-\infty, -\dfrac{67}{10}\right)$

ANSWERS TO FIGURE 3.4

Course: Calculus
Typical Topics you Need to Understand Entering this Class:

- Understanding functions: polynomial and exponential functions.
- 1) What is a function? A function is a relation where for every one input, there is one unique output.

  2) As $x \to \infty$, $f(x) \to \infty$.

  3) $x = -\dfrac{5}{4}$

- Solving logarithmic equations using the properties of logarithms.

  $x = 4$

- A solid understanding of major trigonometric concepts:

  1) $\dfrac{\pi}{6}$

  2) a) the amplitude (3) b) the period (π) c) the horizontal shift (2π)
     d) the vertical shift (1)

  3) $\cot(x)$

  4) $-\dfrac{\sqrt{3}}{2}\sin(x) - \dfrac{1}{2}\cos(x)$

- Finding asymptotes
  Find the equation of the vertical and horizontal asymptotes for the following functions:

  1) Vertical asymptotes are $x = 6$, $x = 2$

     Horizontal asymptote is $y = 0$

  2) Vertical asymptotes are $x = -7$, $x = 4$

     Horizontal asymptotes is $y = 4$

ANSWERS TO FIGURE 3.5

Course: Introduction to Statistics
Typical Topics you Need Entering this Class:

- An understanding of mean, median, mode, and range
  1) Mean: 55.5, Median: 53, Range 82

  2) Mode: 40 and 50

ANSWERS TO FIGURE 3.6

1. 19 R 4
2. 1, 2, 3, 4, 6, 8, 9, 12, 18, 24, 36, 72.
3. 4
4. 40
5. $(1 \cdot 48)$ $(2 \bullet 24)$, $(3 \bullet 16)$, $(4 \bullet 12)$, $(8 \bullet 6)$
6. $(1 \cdot 64)$ $(2 \bullet 32)$, $(4 \bullet 16)$, $(8 \bullet 8)$
7. 7
8. 34
9. 96
10. 15
11. 11
12. One example:
    $50 - 36 \div 6(4) + 16$

Appendix B

Answers to Figure 4.3 and Chapter 4 Practice Problems

1) $7x^2 - 5x - 2$ is an expression because there are a group of terms with no equal sign.
2) For $13x^2 - 9x - 5$,

   a) The coefficients are 13 and –9; the constant is –5.
   b) The terms are $13x^2, -9x, -5$

3) $42x^2 + 35x$

4) $x = 8$

5) example of a monomial- $7x^2y^2$
   example of a polynomial: $9x^2 - 5x - 3$

6) Factors of 84: 1, 2, 3, 4, 6, 7, 12, 14, 21, 28, 42, 84

7) Factors of 96: 1, 2, 3, 4, 6, 8, 12, 16, 24, 32, 48, 96

8) Multiples of 8: 8, 16, 24, 32, 40, 48, 56, 64, 72, 80

9) Multiples of 6: 6, 12, 18, 24, 30, 36, 42, 48, 54, 60

SOLUTIONS TO FIGURE 4.3

Adding, subtracting, multiplying, and dividing fractions.

1) $\dfrac{29}{24}$

2) $\dfrac{1}{30}$

3) $\dfrac{12}{23}$

4) $\dfrac{448}{867}$

Computing multiple operations with fractions:

$$\frac{16}{15}$$

Converting a decimal to a fraction and vice versa.

1) 0.64

2) $\dfrac{7}{8}$

Converting a mixed number to an improper fraction and vice versa.,

1) $7\dfrac{5}{7}$

2) $\dfrac{51}{8}$

Appendix C

Answers to Chapter 5 Practice Problems

Practice with Fractions

1) $\dfrac{1}{15}$

2) $\dfrac{7}{30}$

3) $\dfrac{2}{7}$

4) -1

5) $\dfrac{3x^2 + 17x - 4}{(x-2)(x+5)}$

6) 1

7) $\dfrac{2}{x}$

Practice with Exponents. Simplify the Following:

1) x^{26}

2) y^8

3) y^{40}

4) $4x^4$

5) $\dfrac{z^4}{4y^2}$

6) $\dfrac{1}{x^6}$

7) $\dfrac{1}{y^2}$

Geometry: Angles, Lies, and Rays

1) A ray because it has one end point and extends in one direction.
2) Consider the diagram below:

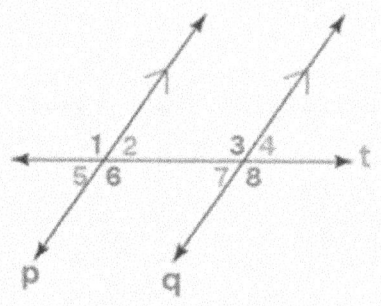

Lines p and q and cut by the transversal, t.

a) ∠1 and ∠2 , ∠5 and, ∠6, ∠3 and ∠4 ,∠7 and ∠8

b) ∠1 and ∠6 , ∠2 and, ∠5, ∠3 and ∠8 ,∠4 and ∠7

c) ∠6 and ∠3 , ∠2 and, ∠7

d) ∠1 and ∠8 , ∠5 and, ∠4

e)

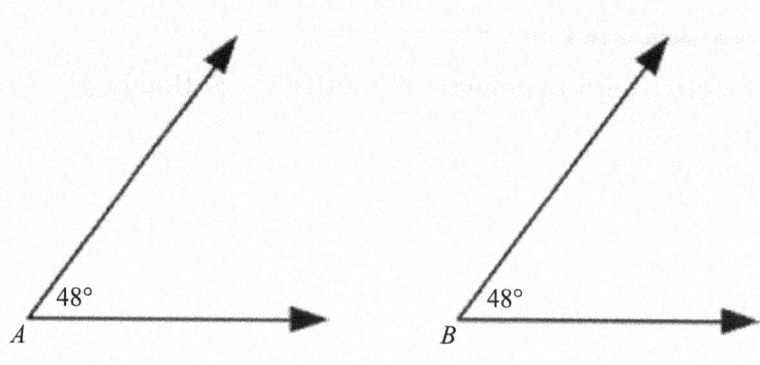

Factoring

Factor the following completely:

1) $5x(x-3)$
2) $3x(3x-7)$
3) $10y^2(1-3y)$
4) $(x+5)(x-9)$
5) $(x+9)(x-8)$
6) $(x+7)(x-12)$

Slope

Find the slope from the following points:

1) $-\dfrac{3}{2}$
2) 1
3) $-\dfrac{1}{3}$
4) $y = 2x - 3$

Appendix D

Answers to Chapter 6 Practice Problems

Basic Algebra Word Problems

1) 15 boxes
2) 10 weeks
3) 20 oatmeal raisin cookies, 58 chocolate chip cookies
4) 3 hours
5) Length is 30 feet; width is 65 feet.
6) 18
7) 93

More Advanced Algebraic Word Problems

1) 4.24 seconds
2) Width is 8 feet; length is 27 feet.
3) The numbers are 10 and 26.

Some Basic Probability

1) $\dfrac{1}{5}$

2) $\dfrac{1}{3}$

3) $\dfrac{1}{4}$

4) $\dfrac{1}{12}$

Index

Note: Figures are indicated by *italics*. Tables are indicated by **bold**.